TRANSACTIONS

OF THE

AMERICAN PHILOSOPHICAL SOCIETY

HELD AT PHILADELPHIA

FOR PROMOTING USEFUL KNOWLEDGE

NEW SERIES—VOLUME 62, PART 6

1972

THE SCIENTIFIC PAPERS OF JAMES LOGAN

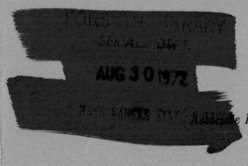

Edited by

ROY N. LOKKEN

Associate Professor of History, East Carolina University

THE AMERICAN PHILOSOPHICAL SOCIETY

INDEPENDENCE SQUARE

PHILADELPHIA

August, 1972

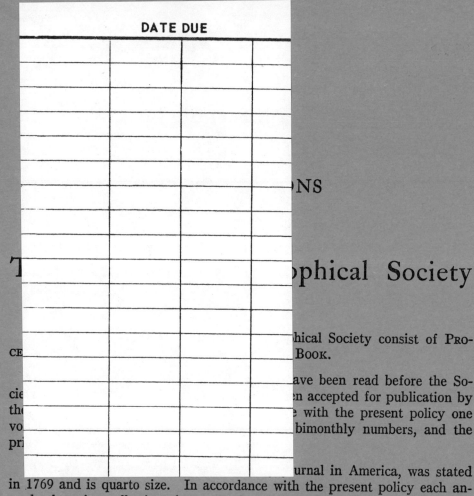

NS

T ophical Society

hical Society consist of PRO-
CE BOOK.

cie ave been read before the So-
th en accepted for publication by
vo e with the present policy one
pr bimonthly numbers, and the

urnal in America, was stated
in 1769 and is quarto size. In accordance with the present policy each annual volume is a collection of monographs, each issued as a part. The current
annual subscription price is $15.00 net per volume. Individual copies of the
TRANSACTIONS are offered for sale. This issue is priced at $4.00.

Each volume of the MEMOIRS is published as a book. The titles cover
the various fields of learning; most of the recent volumes have been historical.
The price of each volume is determind by its size and character.

The YEAR BOOK is of considerable interest to scholars because of the reports on grants for research and to libraries for this reason and because of the
section dealing with the acquisitions of the Library. In addition it contains
the Charter and Laws, and lists of present and former members, and reports
of committees and meetings. The YEAR BOOK is published about April 1 for
the preceding calendar year. The current price is $5.00.

An author desiring to submit a manuscript for publication should send it
to the Editor, George W. Corner, American Philosophical Society, 104 South
Fifth Street, Philadelphia, Pa. 19106.

TRANSACTIONS

OF THE

AMERICAN PHILOSOPHICAL SOCIETY

HELD AT PHILADELPHIA
FOR PROMOTING USEFUL KNOWLEDGE

NEW SERIES—VOLUME 62, PART 6
1972

THE SCIENTIFIC PAPERS OF JAMES LOGAN

Edited by

ROY N. LOKKEN

Associate Professor of History, East Carolina University

THE AMERICAN PHILOSOPHICAL SOCIETY
INDEPENDENCE SQUARE
PHILADELPHIA

August, 1972

Library of Congress Catalog
Card Number 72-76613
International Standard Book Number 0-87169-626-6

PREFACE

ACKNOWLEDGMENTS

In preparing this monograph I have included both James Logan's formal papers, published and unpublished, and some of his relevant personal letters on scientific matters. I have not changed Logan's spelling, but have completed abbreviated words when necessary by inserting omitted letters in brackets. I have also inserted corrections in brackets wherever typographical errors occurred in the original publications.

The directors and staffs of the Historical Society of Pennsylvania, the John Carter Brown Library of Brown University, Library Company of Philadelphia, New York Public Library, University of Wisconsin Library, and the Library of Congress have been very helpful to me in my research.

The English translation of the Latin documents in this monograph, with the exception of *De Plantarum Generatione Experimenta et Meletemata*, is the work of Professor Benedict Monostori of the University of Dallas. Professor Monostori is both a Latin scholar and physicist. I am greatly indebted to him.

I am also indebted to Dirk J. Struik, Donald Fleming, and I. Bernard Cohen, Frederick B. Tolles, Lloyd Lassen, and Louis Bragg for their criticisms and suggestions during the several stages of my work. Dr. Struik's critical reading of the monograph in the last stages of its preparation immeasurably strengthened it.

The University of Texas at Arlington extended several research grants to me which aided me. Dr. John M. Howell, Dean of the Graduate School, East Carolina University, has kindly loaned me the services of several of his typists.

I am much indebted to my wife, Ruth, and to Max Savelle, friend and teacher, and to them I dedicate this monograph.

Every effort has been made to eliminate errors which tend to creep into the transcription of manuscript sources. Such errors as remain in the completed work are mine.

NOTE ON PROVENANCE

Most of the papers and all of the letter books of James Logan are preserved in the Historical Society of Pennsylvania. However, some of Logan's writings, reprinted in this monograph, are to be found elsewhere. The following list defines abbreviations used in this monograph to indicate provenance.

Armistead, *Logan*—Wilson Armistead, *Memoirs of James Logan*, London, 1851

HSP—Historical Society of Pennsylvania

Phil. Trans.—Royal Society of London, *Philosophical Transactions*

Rigaud—Stephen J. Rigaud, ed., *Correspondence of Scientific Men of the Seventeenth Century, including Letters of Barrow, Flamstead, Wallis, and Newton*, 2 v., Oxford, 1841, reprinted by George Olms Verlachsbuchhandlung, Hildesheim, 1965

The John Carter Brown Library, Brown University, provided me with microfilm of their copy of James Logan's *De Radiorum in Superficies Remotius ab Axe Incidentium a Primario Foco Aberrationibus*.

After I had completed this monograph, eight hitherto unknown letter books of James Logan were acquired by the Historical Society of Pennsylvania, whose director immediately restricted them to the exclusive use of Edwin Wolf 2nd, Librarian of the Library Company of Philadelphia, for an indefinite period of time. Mr. Wolf is preparing a detailed catalogue of the library of James Logan preserved at the Library Company of Philadelphia. In preparing the catalogue he proposes to excerpt from the letter books such passages as refer to the books in the library. The catalogue, when completed, will also contain the marginal notes in Logan's books, thereby making them available to scholars in usable form. The catalogue will be a valuable contribution to the intellectual history of eighteenth-century Anglo-America, and it may be used in a future study of Logan's various intellectual interests. Meanwhile the restricted letter books were not made available to me, and it is not clear at this writing what scientific material they contain.

R. N. L.

THE SCIENTIFIC PAPERS OF JAMES LOGAN

Roy N. Lokken

CONTENTS

INTRODUCTION

1.

During the half-century that followed James Logan's arrival in Philadelphia in 1699, Pennsylvania developed from an infant frontier community to the most cosmopolitan, cultured, and progressive province in British America. When Logan debarked from the *Canterbury* in November, 1699, Pennsylvanians were occupied in developing an agricultural-commercial economy and a political system distinctively their own, and they had little time for cultural pursuits. Logan's early life in Pennsylvania reflects the narrowly practical emphasis of everyday living in a frontier community—narrowly practical, that is, except on first days when Quaker settlers observed their fealty to the Divine Being and the Inner Light. As William Penn's

agent, responsible for the property and business interests of the Quaker proprietor, Logan had little time for else but collecting quitrents, managing the land office, and looking after the proprietor's interests in the government of the province. It was not until after Logan had developed a mercantile business, dealing chiefly in skins and furs, that he became active in scientific study and investigations. In this respect Logan followed a trend. Growing wealth and the rise of a class of relatively affluent gentry were accompanied by an increase of cultural life.

Logan's cultural interests actually antedated his arrival in Philadelphia in 1699. They were postponed for some years because of the many practical concerns that occupied his time and attention, and they were revived when his growing affluence enabled him to build a library and to devote more time to intellectual matters. Although Logan had little formal education, he had a sound background in classical learning and mathematics upon which he could build when circumstances permitted.

Logan's only formal education was given him by his father, a university-educated Scottish Quaker who, for a time, kept a Latin school at Lurgan in northern Ireland where James was born in 1674. At Lurgan James Logan learned Latin, Greek, and some Hebrew. Latin and Greek were indispensable to an elementary formal education in Logan's time; further education depended on one's future prospects in the medical, legal, military, and clerical professions. Because the Logans were poor, the professions were unattainable to James. Hence, James Logan was apprenticed to a linen draper in Dublin to learn a trade. The Glorious Revolution of 1688 changed the course of James Logan's life, as it wrought so many changes in the lives of Britishers everywhere. With the army of James II breathing down their necks, the Logans fled from Lurgan, and crossed the Irish Sea to Scotland. Settled in Edinburgh, James Logan taught himself mathematics, and later at Bristol he studied Latin, Greek, Hebrew, French, Italian, and Spanish—all without the aid of a schoolmaster, except that he consulted a French master who helped him with the pronunciation of the Gallic language. For a time Logan taught at a Latin school in Bristol, replacing his father who returned to Lurgan. There was little money in teaching, however, and Logan attempted unsuccessfully to enter the cloth trade in Dublin and Bristol. At Bristol, in 1699, William Penn employed Logan as his secretary, and that September they sailed to Pennsylvania on the *Canterbury*.

During the balance of his life Logan served William Penn and his heirs in a variety of capacities, as Secretary of the Province, Receiver-General, Commissioner of Property, and member of the Provincial Council, to name a few of the offices he held. The foremost defender of proprietary interests in Pennsylvania, Logan was the central figure of a conservative faction which supported proprietary prerogative and represented the wealthier class of Pennsylvanians. For years he was the *bête noire* of the anti-proprietary, middle-class faction which grouped around such popular leaders as David Lloyd. After 1710 Logan developed a successful mercantile business in Philadelphia, dealing chiefly in furs. Investing also in ships, land, iron works, and English securities, Logan accumulated considerable wealth, and built for himself and his family a comfortable mansion, Stenton on Germantown road, and he lived there after 1730. Stenton is still standing.

Logan's scientific activities began, significantly, after his mercantile business was well established and making a profit. Still, his many responsibilities as proprietary agent, public official, political leader, and man of business gave him little time for a life of culture. An accidental fall in 1728 which left him crippled restricted his activities to some extent, and he had more time thereafter for the intellectual life he preferred. Nevertheless, he continued to hold public offices, as a judge and provincial councillor, and, for a time, he governed Pennsylvania as President of the Council. The intellectual life filled his leisure moments, such as they were. Reading, study, philosophical speculation, and scientific research were the entertainment of James Logan and other gentry in the Philadelphia-New York area. Virginia gentlemen may have preferred hunting the fox; Logan was happier resolving a problem in Newtonian calculus.

2.

Between 1715 and 1750 Logan was one of a small group of men in the Philadelphia-New York area who were actively engaged in scientific inquiry and speculation. Among these men were such Crown officers as Colonel Robert Hunter, the royal governor of New York and New Jersey between 1710 and 1719, and William Burnet, royal governor of New York from 1720 to 1728. As the correspondence in this volume shows, Logan delighted in communicating with such well-educated officials in scientific and other intellectual matters. Logan kept in touch also with such intellectually active colonial gentlemen as Cadwallader Colden and James Alexander in New York, and encouraged the scientific studies of young men of lower class who gave evidence of scientific talent—young men such as Thomas Godfrey, John Bartram, and Benjamin Franklin. Logan was older than most, if not all, of the scientific intellectuals in the Middle Colonies with whom he was in contact; he was the best mathematician; and he possessed the best scientific library. For some years he was the leading figure in the group, although he was destined to be overshadowed in the history of early American science by Benjamin Franklin, an ingenious entrepreneur of ideas who was not as learned a scholar as was his Quaker correspondent.

The group of scientific intellectuals in the Middle Colonies with whom Logan communicated was an informal association of men who had similar interests, enjoyed each other's company, and found it convenient to cooperate in working out solutions to scientific problems. Logan never attempted to organize a formal scientific society. Franklin had more interest in and talent for formal organization than had Logan. Aside from local political associations necessitated by his responsibilities as proprietary agent, Logan was not a joiner. He was not associated with Franklin's "Junto," and did not cooperate in organizing the American Philosophical Society. Logan was too much of an individualist, intellectually, to organize "juntos," hold formal meetings, and engage in discussions conducted in accordance with a prepared agenda. He would probably have accepted membership in the Royal Society of London, had it been offered to him, because of the great prestige that was associated with such membership, but he lacked interest in organizing scientific societies on American soil.

Why Logan refused to cooperate with Franklin and Bartram in organizing the American Philosophical Society is a difficult question to answer, aside from the fact that he was not by nature a joiner. One conjecture is that Logan "harbored a grudge against societies of virtuosi because the Royal Society had never made him a fellow. . . ."[1] There is, however, another likely reason.

There appears to have been no communication between Logan and scientific intellectuals in the New England and southern colonies. Upon reading the papers of James Logan, one gets the impression that for him Boston, Williamsburg, and Charleston hardly existed. This significant lacuna in Logan's correspondence does not mean that Logan was provincial. Actually, he was an imperialist who recommended, in 1731, the political and military union of the colonies to protect the British Empire in America from hostile French and Indians in the north and west.[2] This recommendation was drawn from a plan of union suggested by William Penn thirty-five years earlier, and it antedated Franklin's Albany Plan of Union by twenty-three years. Logan's proposal for intercolonial union, however, was entirely political and military, and it was directly related to British foreign policy. Logan evinced little interest in any intercolonial cultural union. He had little regard for the Saints of New England who, he thought, "by their education and institutions [were] naturally and peculiarly stiff."[3] Although Logan, in 1749, blessed Franklin's plan to erect an academy in Philadelphia (an academy which later became the College of Philadelphia and ultimately the University of Pennsylvania), he refused to preside over the organization of a new college at Princeton in New Jersey, in part because he could not approve of a college controlled by Presbyterian clergymen. Logan's papers, moreover, do not indicate any particular awareness of cultural activity in the colonies south of Pennsylvania.

Logan, quite clearly, did not share with his younger contemporaries their interest in the cultural stirrings of provincial America as a whole. Benjamin Franklin and Cadwallader Colden were well acquainted with scientific intellectuals in New England. The American Philosophical Society was founded in Philadelphia to serve all the colonies, and Professor John Winthrop of Harvard was one of the early corresponding members. Cotton Mather's influence on Benjamin Franklin is well known, although it has probably been over-stated. John Bartram in his journeys developed friendly relationships with such southern scientific intellectuals as John Mitchell of Virginia and Dr. Alexander Garden of South Carolina.[4] Franklin and Bartram, and especially Franklin, had a clearer vision of a developing American civilization than did Logan. Logan, in his defense of Thomas Godfrey's claim to the invention of the quadrant with reflecting speculums, expressed a local patriotism which he never developed into anything that remotely resembled an embryonic cultural nationalism. Logan's cultural interests were limited to a small group of friends in the Philadelphia-New York area and scientific intellectuals in England and continental Europe.

3.

Logan's cultural affinity with the Old World is very clear in his scientific papers. For him London and Amsterdam were much closer culturally than Boston and Charleston. He kept in touch with intellectual currents in Europe by corresponding with learned men in England and with London booksellers. Because he was a merchant who either owned ships or had a financial interest in merchant ships, and in any event had business relationships with English merchants, Logan was in continuous contact with the Old World. Transatlantic commerce was a means of communication with virtuosi and professional scientists in England. With cargoes of furs Logan sent orders for books through his London factor and other English contacts, and letters intended for such English scientific intellectuals as William Jones, Peter Collinson, and Sir Hans Sloane.

Transatlantic correspondence in the first half of the eighteenth century was slow, uncertain, and always risky. Letters were carried by captains of merchant ships to their ultimate destination. The lanes of transatlantic commerce were the paths of transmission of scientific ideas and data between English America and Europe. Such transmission was necessarily slow, because a merchant ship normally took about two months to cross the ocean from Philadelphia to London. It was risky, because in wartime (and wartime was the rule in Anglo-French relations in the

[1] Frederick B. Tolles, *James Logan and the Culture of Provincial America* (Library of American Biography, edited by Oscar Handlin, Boston. Little, Brown and Company, 1957), p. 213. Logan had at first suspected that Franklin and his Junto were plotting against the proprietary as tools of Sir William Keith. However, Logan and Franklin were on friendly terms by 1735 (Carl Van Doren, *Benjamin Franklin* (New York, Viking Press, Compass Book edition, 1964), p. 103). There appears to be no evidence that Logan rejected membership in the American Philosophical Society for political reasons.

[2] James Logan, "Of the State of the British Plantations in America: A Memorial" (1732), in Joseph E. Johnson, ed., "A Quaker Imperialist's View of the British Colonies in America: 1732," *Penna. Mag. of Hist. and Biog.* 60 (1936): pp. 97–130.

[3] Quoted in Frederick B. Tolles, *op. cit.*, p. 165.

[4] Michael Kraus, *Intercolonial Aspects of American Culture on the Eve of the Revolution* (New York, Octagon Books, 1964), p. 176.

eighteenth century) a British merchant ship was subject to capture and confiscation by a French man-of-war on the high seas. The imperial postal system, which was developed after 1711, was irregular, and depended entirely on imperial merchant shipping. It was not until after 1755 that significant improvements were made in the postal service between the colonies and the English homeland.[5]

Transatlantic commerce was the means by which cultural contact between the Old World and the New was maintained. Commerce was essential, not only to the expansion of western civilization, but to the maintenance and development of civilization in the colonial communities on the hither edge of the American wilderness.[6] The imperial trade enabled the colonists in English America (transplanted Europeans all) to remain basically Old World (i.e., English, for the most part) in their intellectual outlook—with American variations, to be sure, but basically Old World. The relationship between commerce and the eighteenth-century origins of *The Atlantic Civilization* was pointed out by Michael Kraus in a book which was first published in 1949.[7] Frederick B. Tolles has, in a more specialized study, re-emphasized that relationship.[8] The significance of commerce in the history of early American culture generally, and science specifically, merits further study.

Through his commercial contacts in London and Bristol, Logan kept abreast of scientific developments in the Old World, and he and his intellectual companions in the Middle Colonies exchanged books and information which they received from England. They not only kept themselves informed of scientific progress in Europe, but endeavored to make their own contributions to them. Conversely, scientific intellectuals in England and occasionally on the European continent drew upon Logan and his friends for ideas and data. There was a mutual understanding among virtuosi on both sides of the Atlantic as to the nature of the problems involved in early eighteenth-century science, and the kind of solutions the problems seemed to call for. They understood, for example, that the problem of determining longitude at sea required significant improvements in existing navigational instruments, notably the quadrant. The nearly simultaneous invention of the quadrant with reflecting specula by John Hadley in England and Thomas Godfrey in Pennsylvania was more than a mere coincidence. Their invention was a perfectly logical result of the effort to improve upon Davis's quadrant

so as to develop a better instrument for the determination of latitude and longitude. Logan's interest in the new quadrant as a potentially useful instrument in the solution of problems in positional astronomy was entirely consistent with the trend of thinking among European astronomers at the time. Logan's botanical experiments, also, were consistent with current trends in the science of botany on the European continent.

4.

Like most British scientific intellectuals of his time, Logan was basically a Newtonian. He owned copies of three editions of Newton's *Principia Mathematica*, and taught himself Newton's system of calculus. Books about Newtonian physics and mathematics also filled Logan's bookshelves. Logan found Newton's cosmogony very congenial to his own orderly habits of thought, but, as his correspondence shows (*vide* Part III), he was not a slavish follower of Newton. He did not hesitate to criticize the master when he thought him in error. Logan, moreover, was acquainted with the mathematical work of G. F. W. Leibniz, and credited Leibniz as much as Newton with the discovery of modern differential and integral calculus.

Logan also read Christiaan Huygens's work on optics, and, although he endeavored to render Huygens's demonstrations more intelligible to the lay reader, he very clearly preferred Huygens's geometrical optics to Newton's, although Newton's theory of optics was more widely accepted in the eighteenth century. The works of John Wallis and other European mathematicians were among the books in Logan's library. Many of the ancient mathematicians were represented in Logan's collections and, what is more, he read them.

As the papers in Part I of this volume show, Logan was well read in both ancient and modern sixteenth-, seventeenth-, and eighteenth-century astronomy. He kept abreast of current developments in astronomy; was well acquainted with the work of John Flamsteed, Edmund Halley, and the astronomers on the European continent; and encouraged his friends in the Middle Colonies to make astronomical observations which might be a contribution to scientific progress. Logan, moreover, owned and studied the works of ancient astronomers. Truth being timeless, Pythagoras was as contemporary as Halley.

As was generally true of colonial merchants, Logan had a garden at Stenton in which he delighted to putter when time permitted. He approached botanical science also through his reading in natural religion, and he collected books on agricultural science and botany, such as Richard Bradley's *New Improvements of Planting and Gardening*. Through Peter Collinson, the London Quaker merchant and middleman between scientific intellectuals in the colonies and Europe, Logan was brought into transatlantic communication

---

[5] Michael Kraus, *The Atlantic Civilization: Eighteenth Century Origins* (New York, Russell & Russell, 1961), pp. 26–27.

[6] See Bernard Bailyn, "Communications and Trade: The Atlantic in the Seventeenth Century," *Jour. Economic Hist.* **13** (1953): p. 378.

[7] Michael Kraus, *The Atlantic Civilization: Eighteenth Century Origins*.

[8] Frederick B. Tolles, *Quakers and the Atlantic Culture* (New York, Macmillan, 1960).

with Linnaeus (Karl von Linné), the Swedish botanist who was then in the process of developing a system of classifying plant life. Logan arranged a productive correspondence between his young protégé, John Bartram, and Linnaeus, thereby enabling Bartram to make his contribution to the progress of eighteenth-century botany. Logan assisted Bartram in many ways, and joined him in microscope examinations of the parts of plants. However, as Part IV shows, Logan was more interested in the philosophical implications of his own botanical experiments than in botany as a descriptive science. He was aware of current developments in the other descriptive sciences—biology and medicine, but he took no active interest in them, other than to transcribe articles in the *Philosophical Transactions* which he thought to be of some value.[9]

Logan obtained from England issues of the *Philosophical Transactions* of the Royal Society of London, and read them eagerly as he received them. They were his principal means of keeping up with current trends of scientific thought in the Old World.

5.

Recent historians of early American science have emphasized an American interest in natural history and invention.[10] The emphasis seems all the more attractive since the North American continent offered abundant opportunities for collecting and observations of natural phenomena. Moreover, the practical necessities of living on the hither edge of a vast wilderness nurtured invention. It is true that John Bartram was a collector, rather than a systematic scientist. Bartram made his contributions to the progress of eighteenth-century botany, but it was Linnaeus who developed a system of plant classification. Moreover, Benjamin Franklin, with his stove and lightning rod, was, in that respect, the eighteenth-century exemplar of the American gadget-maker. However, as I. Bernard Cohen has shown, Franklin was more than just a predecessor of Edison; he did have an understanding of theoretical physics which was not beneath that of any European physicist of his time—amateur though he was. Franklin was not the mathematician that Logan was, but his knowledge of physics was good enough to enable him to develop a theory of electricity in advance of its time. Franklin's work on electrical theory gave him an international fame as a scientist

that fitted wonderfully into his later career as a statesman and diplomat.[11]

Logan and the scientific intellectuals in the Middle Colonies with whom he communicated were, for the most part, primarily interested in natural philosophy—astronomy, mathematics, and Newtonian physics. It is easy to write off the theoretical science of these men as naïve, as a recent historian has done, although the thesis that Franklin's success in electrical theory represented "the triumph of naïveté over learning" seems a bit far-fetched.[12] One can accept the fact that they made no earth-shaking contributions to physical science without denying that they at least made an effort. Cadwallader Colden's courageous but naïve attempt to improve upon Newton's mathematical principles by determining the cause of gravitation is an oft-told story. This was a problem which Newton had not himself solved but which he had left to posterity with some suggestions as to a possible solution.[13] Colden's *Principles of Actions in Matter* was not generally accepted by those who read it and expressed their opinion of it. Logan at first thought that Colden's theory had been advanced earlier by Jacques Bernoulli, but subsequently conceded the originality of Colden's speculations.[14] Needless to say, Colden did not solve the problem of the cause of gravitation, but in all fairness it should be added that neither did any European scientist in the eighteenth century.

The natural philosophers in the Middle Colonies were not mere observers and collectors of scientific data; they were also speculative thinkers. That is the real significance of Colden's "extension of the Newtonian principles," and it is also the significance of Logan's work on spherical aberration. Both Newton

---

[9] An example is a manuscript copy of "Observations on a Method of recovering Persons dead in appearance, by distending the Lungs with Air, published in the Last Volume of the Medical Essays &c., printed at Edinburgh 1744 by J. Fothergill M. D.," preserved in Logan Papers, Vol. II, Historical Society of Pennsylvania.

[10] Daniel J. Boorstin, *The Americans: The Colonial Experience* (Vintage Books, New York, Random House, 1964), pp. 162–163, 243–251; Brooke Hindle, *The Pursuit of Science in Revolutionary America, 1753–1789* (Chapel Hill, University of North Carolina, 1956), p. 80.

[11] I. Bernard Cohen, *Franklin and Newton: An Inquiry into Speculative Newtonian Experimental Science and Franklin's Work in Electricity as an Example Thereof* (Philadelphia, Mem. Amer. Philos. Soc. **43** (1956)), pp. 36–38.

[12] Boorstin, *op. cit.*, p. 252. In a bibliographical note on p. 407 Boorstin rejects the conclusions reached by Professor Cohen in his *Franklin and Newton*. However, this writer feels that Cohen's conclusions are based on more systematic research in the original sources than are Boorstin's. William Appleman Williams, in his *The Contours of American History* (Chicago, Quadrangle Paperbacks, 1966), p. 92, also refers to Franklin as "only" an amateur. As Dirk J. Struik has observed in a note to this writer, "It was still a good time for the amateur."

[13] See Sir Isaac Newton, *Opticks: or, a Treatise of the Reflections, Refractions, Inflections & Colours of Light*, "Queries" and "General Scholium." Also, see Brooke Hindle, "Cadwallader Colden's Extension of the Newtonian Principles," *William and Mary Quart.*, 3rd ser., 13 (October, 1956): p. 463. The Bernoullis and Christiaan Huygens had made earlier attempts to determine the cause of gravitation. See Brooke Hindle, *The Pursuit of Science in Revolutionary America*, p. 44; A. Pannekoek, *A History of Astronomy* (New York, Interscience Publishers, 1966), p. 261.

[14] Franklin to Colden, Philadelphia, October 16, 1746, in Leonard W. Labaree, ed., *Papers of Benjamin Franklin* **3** (New Haven, Yale University Press, 1961): p. 90. Logan appears to have referred to Jacques Bernoulli (1654–1705), author of *Dissertatio de Gravitate Aetheris* (Amsterdam, 1683).

and Huygens had denied the possibility of proving the laws of spherical aberration with absolute mathematical rigor. Logan courageously set out to do the impossible, and his paper on the subject was published in the Netherlands. The paper attracted little attention; to the modern reader Logan's mathematical proofs appear tedious and somewhat obscure.

It is ironical that Logan's most original contributions were in botany—a field of science in which he had not been directly interested until he read William Wollaston's *The Religion of Nature Delineated*. His botanical work arose from his interest in a philosophical problem more profound and fundamental than the sexual generation of plants. Basically, in all his scientific studies and investigations Logan was a philosopher who sought the foundations of reality and the meaning of existence—all verifiable by observation and experiment and, in the case of physical theory, subject to rigorous mathematical proof.

6.

Abbé Raynal's charge, in 1774, that "America has not yet produced one good poet, one able mathematician, one man of genius in a single art or a single science" has often been quoted by historians, if only to restate Thomas Jefferson's refutation. Environmental conditions existing in America were obviously not conducive to the production of scientific giants. Jefferson, of course, reminded European detractors of America that Franklin and David Rittenhouse were among the scientific geniuses of the age.[15] A recent historian has suggested that the work of Franklin and Rittenhouse exemplified "the limits of American culture in the colonial age," and attributes those limits to Americans' "democratic way of thinking."[16]

A fact frequently neglected by historians of early American science, however, is that the intellectual climate throughout western civilization had changed during the eighteenth century. The great days of Galileo, Kepler, Newton, and Leibniz were over; the day of Darwin and Helmholtz was yet to come. In the early eighteenth century the scientific revolution of the three preceding centuries settled down to a period of professional work on technical problems, the invention of gadgets, and efforts to apply scientific rationalism to political and social thought. The emphasis after Newton was, in part, on putting into practical use the great scientific theories developed during the centuries immediately preceding. The Greenwich Observatory had been created late in the seventeenth century for a practical purpose—the dis-

covery of a foolproof method of determining longitude at sea. The observations of the moon's place by John Flamsteed and Edmund Halley were to serve that practical purpose—as well as to contribute to the discovery of scientific truth. Benjamin Franklin's electrical experiments were inspired by activity in the field among European scientists. Eighteenth-century progress in the knowledge of electricity owed as much to Charles François de Cisternay Du Fay and Charles Augustin Coulomb as it did to Franklin.[17] Franklin's contributions earned a notable chapter in the history of modern science, but, unless Franklin's work is placed in its proper historical perspective, the chapter is poorly written. The success of Franklin's experiments is attributable to his talent as a scientist; it is no indication of anything uniquely American in the first half of the eighteenth century. Franklin's electrical experiment led to the invention of the lightning rod, but if Franklin was a gadget-maker, so were Huygens, Papin, Newcomen, Savery, and James Watt in Europe.[18] Logan was not an inventor, but his letters evince an interest in gadget-making in the Old World.[19]

Moreover, during the eighteenth century there was a pronounced trend toward the professionalization of science. The sciences became too technical for significant contributions by laymen. Scientific education spread, and scientific experiment, observation, and invention shifted from the home workshop to the university classroom, laboratory, and observatory. In America there was still an important laity in the sciences (witness Logan, Colden, and Franklin), but even there a trend toward academic professionalization was evident (witness John Winthrop of Harvard). However, during the period of Logan's scientific labors the day of the great lay scientists was almost over.

Of course it is true that, in any event, environmental conditions in America would have remained a deterrent to the rise of anyone like Newton on American soil. The scientific intellectual in the Middle Colonies was a busy man. He was a public official whose days were well filled performing the duties of office. He was a merchant whose time was necessarily devoted to business, if he was to pay his debts and provide for his family. James Logan was both public official and merchant, which makes it all the more remarkable that he could find time at all for scientific inquiries. That was the paradox of this remarkable colonial virtuoso. Being both merchant and public official was at once an advantage and a disadvantage. It

---

[15] Thomas Jefferson, *Notes on Virginia* (Harper Torchbooks, New York, Harper & Row, 1964), pp. 64–65.

[16] Boorstin, *op. cit.*, p. 244. Of course it does violence to the intellectual history of pre-Revolutionary America to attribute to Logan or any of his contemporaries a "democratic way of thinking."

[17] René Taton, ed., *History of Science: The Beginnings of Modern Science from 1430 to 1800*, translated by A. J. Pomerans (New York, Basic Books, 1964), pp. 472–488.

[18] *Ibid.*, p. 470. Also see Dirk J. Struik, "The Determination of Longitude at Sea," *Actes du XIᵉ congrès international d' histoire des sciences*, 4 (1966): pp. 262–271.

[19] See, for example, Logan to Hunter, Philadelphia, April 2, 1719, Logan Letter Books, II, 205; Logan to Hunter, Philadelphia, April 23, 1719, *ibid.*, II, 208.

facilitated intellectual intercourse with men of like interests, but it limited one's time for intellectual activity. This was Logan's recurrent complaint, especially as he grew older and felt himself declining in physical strength and intellectual power.

Nevertheless, Logan achieved a prodigious amount of reading and study in spite of the heavy demands on his time. He wrote papers, and published some of them in England and the Netherlands. He instructed other scientific intellectuals in the Philadelphia-New York area, and patronized younger, humbler men who, under his guidance, made contributions of their own to the science of their time—Thomas Godfrey, John Bartram, and Benjamin Franklin. None of them were Newtons, but then there were no Newtons in eighteenth-century Europe either.

There is a tendency on the part of some historians of the colonial period to write off colonial culture as derivative and imitative, because it was more European than American.[20] The very word "derivative" applied to the culture of colonial America before 1763, however, indicates a failure to place that culture in its proper historical frame of reference. That frame of reference was European, not American. Recent computerized scholarship has indicated very strongly that there was no significantly conscious American nationalism until after 1763.[21] Logan was loyal enough a Pennsylvanian to defend the claims of Thomas Godfrey, but he was primarily and basically a European. His cultural interests, his thought, his learning, and his taste were European, as were those of his friends in the Philadelphia-New York area. The history of colonial culture will be more clearly understood when it is seen as part of the history of European culture. Nationalistic pressures were one day to bring about a more Americanized variant of western culture, but that day was not to come until long after the death of James Logan.

7.

Science and religion were more closely related in the eighteenth-century European mind than they are today. The religious ethos of the seventeenth century, to be sure, was eroded by the secularizing influence of the Age of Enlightenment. Religion was modified by scientific rationalism, but it was not eradicated. The scientific intellectuals of the period did not ignore Deity in studying His Creation. Newton devoted most of his life to theological studies, rather than to the mathematics and physics which made him the most significant scientist of the period. Enlightenment intellectuals, being whole men, sought to understand all things in their relations—God, man, the universe, number, the natural world of living things. Logan, who was an Enlightenment intellectual and a whole man, sought to make all knowledge his province, and above all he sought to discover those relations which integrate knowledge and disclose the underlying unity which defines God's Creation.

In Logan's mind there was no conflict between science and religion. Science, in fact, supported and strengthened religion, for through scientific inquiry one learned more fully the handiwork of God. In the study of the natural universe one observed a cosmic order governed by Reason. Logan was inspired by the harmony prevailing throughout the world of nature and the logical relationships which pervaded the whole of Creation, all ascribable to the Supreme Being, the Author of All. All the beauties of Nature, the "handmaid of God," as Logan called her, "lie open to the view and consideration of us all and they are the just objects of our admiration."[22] Winthrop's "city on a hill," a godly society devised by Puritans in the New World in covenant with God as a model for all men to follow, became in the eighteenth century a "heavenly city," a God-given model of natural harmony and reason which all men could profitably imitate in their personal conduct and social order. Scientific inquiry not only gave Logan fresh reason to worship the Deity who was the Creator of such a wonderfully harmonious and rational cosmos, but provided him with a world-view upon which he based his ethical and political thought.[23]

8.

As a scientific intellectual Logan was a stickler for accuracy. He was quick to pounce on an error, even when the error was made by such an intellectual colossus as Newton. So it was that he insisted on accuracy on the part of the members of the Philadelphia-New York group in their reports of astronomical observations, their mathematics, and other scientific investigations. Logan could at times be severe in his insistence on correctness of fact and scientific procedure. He could be severe with himself, too, because he occasionally erred, and was quick to admit it. There was nothing vain about Logan. His learning was tempered by a studied humility and intellectual modesty. Like John Locke, he did not presume too much. Knowledge was necessarily limited, and it was human to err. However great the mind, it could

---

[20] See, e.g., Clinton Rossiter, *Seedtime of the Republic: The Origin of the American Tradition of Political Liberty* (New York, Harcourt, Brace and Company, 1953), p. 119.

[21] See Richard L. Merritt, *Symbols of American Community, 1735–1775* (New Haven, Yale University Press, 1966), especially ch. 8.

[22] James Logan, "The Charge Delivered from the Bench to the Grand Inquest, At a Court of Oyer and Terminer, and General Jail Delivery, Held for the City and County of Philadelphia, April 13, 1736," in Wilson Armistead, ed., *Memoirs of James Logan* (London, 1851), pp. 122–123.

[23] Roy N. Lokken, "The Social Thought of James Logan," *William and Mary Quart.*, 3rd ser., **27** (January 1970): pp. 68–89. See also Paul W. Conner, *Poor Richard's Politicks: Benjamin Franklin and His New American Order* (New York, Oxford University Press, Galaxy Book, 1969), pp. 173 ff.

never penetrate the screen behind which nature hides her deepest mysteries. But a learned mind is a disciplined mind. Error in the employment of known and knowable facts is difficult to justify. Worse yet, from the point of view of James Logan, was error in ethical judgment. He was most critical of Newton when it appeared to him that the great mathematician had attempted to deceive his readers. Logan was very disturbed when the third edition of *Principia Mathematica* contained a portrait of Newton when he was much younger; such deception seemed to be entirely beyond forgiveness.

Logan did not identify intellectual aptitude with pecuniary wealth and social status. Whatever is knowable in the external world may be discerned by talented men, whatever their outward circumstances. Social station and economic wealth are not necessarily keys to discovery and invention, although they may help in providing books and leisure. It was Logan's awareness that there is no necessary correlation between genius and aristocratic pretensions that led him to regard Thomas Godfrey, a humble glazier, with intellectual respect. Logan had as much confidence in Godfrey's self-taught mathematics as he had in his own, sometimes even more. Logan's respect for Godfrey and Bartram as scientific intellectuals indicates that in the life of the mind Logan regarded social class distinctions as irrelevant. To be sure, he was never a social equalitarian. He had little liking for the common sort of people, as such. But in intellectual pursuits he gave full recognition to merit, even when merit wore common dress.

Personal correspondence forms the bulk of James Logan's literary remains. Logan was a prolific letter writer, and his letters tended to be very long. As the letters in this volume show, his literary style was somewhat involved. Unlike Franklin he did not try to emulate a polished writer like Joseph Addison. Logan wrote at white heat, and jotted down everything that was on his mind, and his mind was always full of all sorts of information and ideas. For that reason Logan's letter books are encyclopedic in content. The researcher interested only in a single topic has the problem of analyzing each letter in search of relevant material. But the historian who seeks a coherent study of the life and thought of early eighteenth-century English America will always be grateful.

I. ASTRONOMY AND ASTRONOMICAL INSTRUMENTS

Logan wrote no formal published work on his astronomical observations, but his correspondence and unpublished pieces reveal an encyclopedic knowledge of the astronomical scholarship of his time. He had an intense interest in the subject, which extended throughout most of his life and which transcended the mere practical necessity of determining latitude so as to locate with certainty the boundary lines between neighboring colonies. Logan, of course, used his knowledge of astronomy in studying the problem of the Maryland-Pennsylvania boundary line (a problem solved long after his death), but the following letters show quite clearly that Logan loved the study of astronomy for its own sake. He was well versed in the progress of astronomical knowledge during the period of the scientific revolution and in the work being done by the astronomers of the early eighteenth century. Logan knew probably more about astronomy than any other person in the Middle Colonies before 1750. His letters reveal him as the tutor of other intellectuals in the Middle Colonies who followed his leadership in astronomical observations.

Logan's private library contained the astronomical works of both the ancient and modern authors, including Ptolemy, Copernicus, Tycho Brahe, Galileo, Kepler, and John Flamsteed.[1] According to Dirk J. Struik, "Logan, as an admirer of Newton, was a Copernican, something still slightly controversial in his day," but probably not among the scientific intellectuals in English America.[2]

Although Logan's interest in astronomy extended throughout most of his life, he was most actively engaged in astronomical observations between 1717 and 1720. Early in 1717 Logan moved to a house on Second Street in Philadelphia which provided him an opportunity to construct a small observatory. Streets in Philadelphia ran nearly north and south, and most houses had no windows facing in those directions. Logan's new home, however, had "a more convenient backside" where an observatory could be installed. He wrote to a clockmaker in England for telescope lenses and an astronomical clock for the projected observatory. Whether or not he built the observatory is not clear, but he did use a telescope and a quadrant in making astronomical observations during this period. For a time he used a quadrant owned by Sir William Keith, then the deputy governor of Pennsylvania; Cadwallader Colden, a scientific intellectual and government official in New York, whom Logan had met when he was a physician in Philadelphia, subsequently borrowed the quadrant for a while. Working with such instruments caused Logan to become aware that accurate observations depended on the efficacy of the instruments used, and his study of the instruments made him acquainted with dioptrics and the problem of atmospheric refraction. A close examination of the problems of refraction and dioptrics resulted in separate studies which are reprinted and discussed in Part II of this book.

---

[1] Frederick B. Tolles, *Meeting House and Counting House: The Quaker Merchants of Colonial Philadelphia 1682–1783* (New York, W. W. Norton Co., The Norton Library, 1963), p. 182.

[2] Dirk J. Struik to this writer, Cambridge, Mass., February 7, 1969.

As the following correspondence shows, Logan in 1717–1720 was primarily interested in the theoretical aspects of astronomical science. His intentions in 1717 were to attempt a solution of the problem of the solar parallax and to test Newton's theory of the moon's place. As his letters show, Logan was primarily concerned about positional astronomy and the determination of distances between stellar bodies essential to a solution of the more significant problem of the size of the universe. The determination of the solar parallax was a question that had long baffled the greatest astronomers of Europe. Astronomers had been attempting to estimate the distance between the sun and earth since ancient times. In 1672 J. D. Cassini in Paris concluded, on the basis of observations of Mars made by Jean Richer at Cayenne in South America, that the solar parallax was of the order $9\frac{1}{2}''$, corresponding to a distance of about 87,000,000 miles.[3] In 1676 Edmund Halley's observations of the transit of Mercury gave him the idea that observations of the transit of Venus would result in a more accurate determination of the solar parallax. A Mercury transit is not very useful for this purpose; it lasts only a few minutes—not enough time for meaningful observations that would yield the distance of the sun from the earth. A transit of Venus takes several hours, and is therefore generally satisfactory for such observations. In 1691 Halley predicted that the next transit of Venus would take place in May, 1761, and in 1716 he outlined a plan for observations of the transit of Venus and determinations of the solar parallax, although he knew that he would not live to take part in them.[4] Halley's 1716 article in the *Philosophical Transactions* created a stir in the learned world, and this is probably the reason for Logan's interest in the problem.

In his letter to Joseph Williamson in March, 1716/17, Logan wrote that he had "lately thought of a method of taking the Altitude of a Star within a very few seconds viz 4 or 5 at most with no great change or trouble, yet with the utmost exactness. . . ." He proposed to use this method in an effort to determine the solar parallax. It is not clear what Logan's method was, and there is no evidence that he met with any kind of success in his experiments.

Observations of the transit of Venus in 1761 and 1769 were made in widely separated parts of the world, giving various figures for the solar parallax—from 8.55″ to 8.88″. The atmosphere around Venus made it difficult to determine exactly the moment of Venus's contact with the sun's disk, hence the divergence in the determinations of the values of the solar parallax. The observations of 1761 and 1769, however, represented significant progress toward an eventual solution of the problem. Years later, Laplace

derived from his mathematical study of the moon's motion a solar parallax of 8.6″. Observations of the transit of Venus in 1874 yielded figures for the solar parallax varying from 8.76″ to 8.88 ± 0.04″. The solar parallax is now determined to be 8.80″ "with the certainty that it is accurate to some thousandths of a second, 1/3000 of its amount."[5]

In the early months of 1717 Logan for the first time calculated the moon's place by Newton's theory. In this experiment he attempted to use William Whiston's explanations of Newton's theory of the moon, published in *Astronomical Praelections*, and, disappointed in them, concluded that Whiston's explanations were of no value. What astonished Logan was his discovery that Newton and John Flamsteed had differed significantly in their methods of finding the eccentricity of the moon's orbit. The fact is that, while Newton's theory was frequently tested by both professional and amateur astronomers, it was purely a theory which did not altogether fit the observed facts. Lunar motion has always been a complicated problem in positional astronomy, and the establishment of reasonably accurate tables of the moon has long been a highly important activity in the field. Because of the inadequacies of Newton's theory, Halley addressed himself to the task of making observations of lunar motion and of compiling tables of the moon's place. Tables of nearly 200 lunar observations were added as an appendix to the second edition of Thomas Streete's *Astronomia Carolina*, published in 1710. Logan had a copy of this edition of Streete; in his correspondence he referred to the appendix as the "Caroline Tables." Halley compiled a complete set of lunar tables in 1719; although they were printed, they were not published until 1749. When Logan was in London in 1723, he found printed sheets of the lunar tables in the bookstore of William Innys, a bookseller and publisher. Logan subsequently wrote that Innys "secretly sold them to me in sheets at a price not at all modest. . . ."[6] Logan brought the tables back to Philadelphia with him, studied them, entered annotations, and prepared a critical analysis of them in Latin. This analysis, "Calculus motus lunae ex his tabulis halleianis," was never published. The original manuscript is preserved at the Library Company of Philadelphia. The "Calculus motus lunae" contains astronomical notes, explanations, and observations which, according to Edwin Wolf 2d, "vastly enriched" Halley's tables.

---

[3] A. Wolf, *A History of Science, Technology, and Philosophy in the 16th and 17th Centuries* (New York, Macmillan, 1935), p. 175.
[4] *Ibid.*, p. 184.

[5] E. G. R. Taylor, *The Mathematical Practitioners of Hanoverian England, 1714–1840* (Cambridge, University Press, 1966), p. 44; A. Pannekoek, *A History of Astronomy* (New York, Interscience Publishers, 1961), pp. 286–287, 345, 348.
[6] *Calculus motus lunae ex his halleianis.* According to Edwin Wolf 2d, Logan met Halley at Innys' bookstore, and purchased the sheets of the lunar tables for a guinea. (*The Annual Report of the Library Company of Philadelphia for the Year 1955*, p. 23.) That may be true, but Logan did not mention in his *Calculus motus lunae* that he had met Halley at Innys' bookstore.

Logan's manuscript contains many specific references to Halley's tables. Hence, it cannot readily be understood, even in English translation, without the tables. Since Halley's tables are 210 pages long, it is not feasible to include either them or the "Calculus motus lunae" in this volume.

Lunar theory was of central importance in Logan's astronomical studies, both because of its relevance to the problem of determining the solar parallax and because, as the sequel will show, of its importance in determining longitude at sea. However, Logan, as was true of other scientific intellectuals in the middle colonies, was also interested in determining the latitude and longitude of the stars. He had in his possession published tables of the stars' places, and he tested them by observation. In quest of accuracy he compared his own observations with those of other members of the New York-Philadelphia group of scientific intellectuals.

The determination of latitude in the New York-Philadelphia region also concerned Logan, not only as a practical matter involving the location of boundary lines but as a matter of observational accuracy in determining the places of the stars. Logan believed that "the only sure method of finding the Latitude" was by the pole star, depending, of course, on the accuracy of the instrument used. In a letter to Governor Robert Hunter of New York, dated September 2, 1718, Logan recommended that observations be taken about December 20 of the pole star's greatest altitude above the pole and its least altitude under the pole. He suggested observations about 6 A.M. and 6 P.M. of the same day, repeated observations before and after that day, "and the half of these two added together will be the true Latitude." He warned, however, that allowance would have to be made for atmospheric refraction.

Logan preferred the pole star in determining the latitude because of the slowness of its motion. However, he was fully aware of the problems of positional astronomy in his time. He warned Hunter, in the letter cited above, that the declination of the pole star was somewhat uncertain. Moreover, the right ascension of the pole star was also difficult to ascertain, and the determination of the longitude of the pole star depended on knowledge of its right ascension. Knowledge of the right ascension and longitude of the pole star was essential in determining its declination. In spite of these difficulties Logan regarded the pole star as the most stable point in the heavens for the determination of latitude.

Astronomical observations made Logan aware of the importance of instruments in arriving at accurate determinations. The experience he had with Governor Keith's quadrant in 1718 caused him to suspect that observations made with defective instruments resulted in great errors. The invention of a new quadrant with reflecting specula interested Logan, therefore, not merely as a new and better navigational instrument but as a potentially better instrument for astronomical observations.

The new quadrant was invented at about the same time by an Englishman, John Hadley, and a Philadelphian, Thomas Godfrey—both unknown to each other and working independently.[7] Hadley ranked just below Halley and Flamsteed as a prominent English astronomer. Godfrey, by contrast, was a young glazier, of limited means and almost entirely self-taught. Logan, who normally preferred the company of upper-class people, was attracted to Godfrey because of the young man's easy mastery of Latin and his quick understanding of Newton's *Principia Mathematica*. When Godfrey invented the new quadrant, Logan, who did not then know of Hadley's similar accomplishment, was so impressed that he wrote a description of it in a letter to Edmund Halley, the Secretary of the Royal Society.

The result was a controversy in which apparently unwarranted charges were made on both sides. Halley evidently thought it presumptuous that such a common person as Godfrey could have invented such an instrument. Logan's countercharge that Hadley had borrowed the idea of a quadrant with the reflecting specula indirectly from Godfrey by way of English sea captains who knew of Godfrey's invention appears to be without merit. Halley, of course, was defending the claims of one of the prominent English scientists of the time. On Logan's part there is an element of local patriotism in his spirited defense of Godfrey's rights as first inventor. But Godfrey had his defenders in England, too. In 1734 William Jones, the mathematician and a correspondent of Logan, appeared before the Royal Society of London, and argued the case for Godfrey as at least one of the inventors of the new quadrant. Peter Collinson also assisted Godfrey by having Logan's detailed description of the invention complete with drawings published in the *Philosophical Transactions*. Subsequently, the Royal Society decided in favor of both Hadley and Godfrey as co-inventors, and each was awarded £200 for his services to navigation. Somehow the Society got the impression that Godfrey was a heavy drinker, and voted to give him a clock worth £200 in place of the money. The instrument, then called "Hadley's quadrant," came to be known as the "sea-octant," and about 1747 Captain John Campbell used it to

---

[7] Actually Sir Isaac Newton had invented such an instrument about 1700, but he had not published it and it had not become known. Even earlier, in 1666, Robert Hooke had made and described a somewhat similar reflecting instrument, although it was inferior to the quadrant (or octant, to be more accurate) devised by Hadley and Godfrey. (See A. Wolf, *A History of Science, Technology, and Philosophy in the 18th Century* (2 v., New York, Harper & Row, Harper Torchbooks, 1961) 1, p. 146; A. Pannekoek, *A History of Astronomy*, p. 291.)

measure lunar distances. The octant was subsequently modified and became our present-day sextant.[8]

In the history of science it matters little who invented the octant first. What is more important is the reason for the invention. According to Logan, Godfrey's primary interest was in developing a better navigational instrument for use in seagoing commerce. However, in his letter to Halley, dated May 25, 1732, Logan expressed greater interest in the astronomical uses of the new quadrant, although, to be sure, the astronomical problem he had in mind was related to navigation at sea. There had long been a problem in determining longitude, especially at sea. Galileo had puzzled over it. The Greenwich Observatory was erected, by order of Charles II, for the principal purpose of discovering a practicable method of determining longitude. Newton advanced the idea that a table of the moon's motions would provide a sort of astronomical clock which could be used in determining longitude at sea.[9] John Flamsteed, the first Royal Astronomer, worked on a table of the moon's motion. Newton tried to work out a table of the predicted motions of the moon, but found it too difficult. Halley was at work on this problem in Logan's time. In 1714 the problem had not yet been solved, and British sea captains, disturbed by the delay, submitted a petition to the House of Commons, requesting that a solution of the longitudinal problem be expedited. Consequently, the British government offered a reward to anyone making a significant contribution to a solution of the problem. It was not until after Logan's death that such rewards were finally given—to Leonard Euler who developed new mathematical techniques on the basis of which reasonably accurate calculations of the moon could be made, Tobias Mayer who made such calculations and published his tables in 1752, and John Harrison who invented the marine chronometer.

In his 1732 letter to Halley, Logan suggested that the new quadrant might be a valuable instrument for the determination of the future motions of the moon, and that would be the basis for the determination of longitude. It seems clear on the basis of Logan's letters to Halley and William Jones that his only purpose was to be helpful in the efforts to resolve a difficult problem in astronomy. There is no evidence that

Logan sought anything for himself. He hoped, of course, that Godfrey would benefit, for he believed the young man to be deserving of success. But Halley's charge that Logan was an impostor and a fraud was without foundation.

A side product of Logan's astronomical observations was his interest in the problems of atmospheric refraction. That interest involved the phenomena of lightning in thunderstorms and the appearance of the sun and moon to the observer on earth at dawn and sunset. Upon reading Stephen Hales's *Statical Essays*, Logan found a passage which indicated that the crooked or angular appearance of lightning was a phenomenon for which scientists had not yet found an explanation. In a letter to Sir Hans Sloane, Logan suggested that atmospheric refraction explained the phenomenon. The relevant extract from the letter was published in the *Philosophical Transactions*.

Another phenomenon which puzzled the scientific intellectuals of Logan's time was "the Sun's and Moon's appearing so much larger at rising and setting, than when in a greater Altitude. . . ." Logan explained the appearance as an illusion caused by atmospheric refraction. The explanation was not new, as Logan confessed; it had been suggested by others. However, Logan described observations which he thought would be a contribution to the subject. The portion of a letter from Logan to Sir Hans Sloane dealing with this subject was also published in the *Philosophical Transactions*.

Atmospheric refraction was long an annoying source of error in astronomical measurements, and it goes back to Ptolemy. Ptolemy measured the refraction of light rays passing from air through glass and water, and observed that the angle of refraction was always less than the angle of incidence. This observation led him to conclude that there was a difference between the apparent position of a star and its real position because of refraction by the atmosphere. His assumption that the proportion between the angle of refraction and the angle of incidence was constant was, however, erroneous. Bernhard Walther (1430–1504), an astronomer in Nuremberg, in 1488 explained the upward displacement of the sun near the horizon as due to atmospheric refraction. In about 1620 Willebrord Snell discovered the law of refraction $\sin i/\sin n = n$; this sine law of refraction was first published by Descartes in 1638. It was Sir Isaac Newton who pointed out that refraction is affected by variable factors, temperature and air pressure. Logan's observations are in accord with the present-day definition of atmospheric refraction as "the bending of a beam of light (or other wave ray) as it travels through layers of air of varied density."[10] Logan also suggested the observer as a variable subjective factor.

[8] Carl and Jessica Bridenbaugh, *Rebels and Gentlemen: Philadelphia in the Age of Franklin* (New York, Oxford University Press, 1965, Galaxy Book paperback), pp. 307–308; Dirk J. Struik, "The Determination of Longitude at Sea," *Actes du XIᵉ congrès international d'histoire des sciences*, 4 (1960); p. 270; A. Wolf, *A History of Science, Technology, and Philosophy in the 18th Century* (2 v., New York, Harper Torchbooks, 1961) 1: p. 149, 151; Henri Michel, *Scientific Instruments in Art and History*, translated by R. E. W. Maddison and Francis R. Maddison (New York, Viking Press, 1967), p. 61.

[9] The idea, of course, was not new, but it had never been successfully worked out. (See A. Pannekoek, *A History of Astronomy*, pp. 276–277.)

[10] *The Harper Encyclopedia of Science*, ed. by James R. Newman (4 v., New York, Harper & Row, 1963) 1: p. 112.

Atmospheric refraction was an important scientific problem in the early eighteenth century, but Logan gave greater attention to the problem of refraction through telescopic lenses. A general study of that problem resulted in two of Logan's major scientific writings, both in the science of optics.

LOGAN TO JOSEPH WILLIAMSON,[11] PHILADELPHIA, MARCH 1, 1716/17, LETTER BOOKS OF JAMES LOGAN, Vol. II, p. 155, HSP.

· · · · · · · · · · · · · · · ·

I have long disappointed thee in the Observations I was to make. Not only too much business of a different kind, but the want of conveniencies where I lived prevented me, for the principal Streets of this town lie Nearly North & South & therefore have no Windows to either, but I am now removed to a House w[i]th a more convenient backside where I propose to build a Small observatory, and have faln on a way to take a Meridian Line to a minute in Space, that is to 4 Seconds of time in the Sun's motion. I could doe it with yet greater exactness but it must be with more trouble. The best method I find is when the Pole Star is near, that is a very little above the 6 hour line, for then its easting and Westing is for several Minutes in time imperceptible, at that time drawing a line by the Star's place of a Sufficient length the Stars distance from the Pole may be sett off w[i]th great exactness by a tan[gen]t w[hi]ch will give the point of a true Meridian from the place of the eye at the first observation.

But I am in great want of Glasses. I wish therefore thou wouldst speak to my friend Wilson[12] and if he will afford them reasonably procure of him some Tellescope Glasses of several Sizes as well Object as eye Glasses. I should be glad to have out of the largest focus that he has a disk be made for they generally charge those the highest, yet when the disk is once made the Glasses are easiest wrought off in them & in much less time than those of a greater convexity. I know he frequently makes these of 15 foot focus but if I had one of 25 or 30 feet I could easily mount it. I would have one of every Size at 3, 6, 10, 15 feet & upwards and Eye Glasses of 3, 4, 6, or 8 Inches proportionably. I should desire also a small object glass or two not above the breadth of a shilling or rather less of 12, 18, & 24 Inches w[hi]ch eyeglasses proportionable for certain uses I should find for them if the prices be reasonable, I would take the most of them. I know these latter are cheap viz about a shilling or 18d a piece, pray know the prices (he warranting

them to be good) of the other Sorts. perhaps if his be high as good may be had of more of the young men who about 7 years agoe sett up together in Pauls Ch[urch] Yard, & one of them afterwards removed into Cornhills, but I would have good ones. Wilson I suppose would Sell his Object & eye Glasses for his 16 foot Tellescopes for 10/ or under, p[er]haps half the money.

If I build the place I have mentioned I must have another Clock for it but shall pay thee the full price, for an abatem[en]t in one was too much unless I had answered expectations better. I would have it a right good one of the Same kind with the other, that is w[i]th a Spiral for the Striking part, & I should desire the Pendulum exactly Sett to equal time there for some Months that I might observe the difference in this Latitude.

Having lately thought of a method of taking the Altitude of a Star within a very few seconds viz 4 or 5 at most with no great charge or trouble, yet w[i]th the utmost exactness, I have a Strong Inclination to make some experiments about the Parallax of our Great Orb viz the Earths about the Sun. Our good friend J fflamsted has found the difference of the Altit[ude] of some Stars at about 6 Months distance to be about 45″ w[hi]ch W Whiston[13] will have to be that parallax & will admitt of no contradiction Positiveness having ever been that unhappy man's greatest misfortune. But I find D Gregory[14] is rather of opinion that this difference is owing to the Nutation of the Earth in its Orbit according to Sr Isaac's Principles. I find also that the ingenious J. fflamstead himself is very doubtfull in the case & cannot with any assurance assign that Parallax to be the cause, and indeed there is this Strong Objection ag[ains]t it, that as there is a great probability that the fixed Stars are distant in some measure in a reciprocal ratio to their apparent magnitude, So if this difference of Altitudes were owing to the Parallax of the Orb it must be greater in relation to the greater or nearer Stars & less w[i]th the more remote, but I find the observations that have been made produce nearly the same difference as to Stars of all magnitudes[.] pray discourse J. ffl[amsted] if in health about it. I have never Seen his Lett[e]r on that Subject published in

----

[11] Joseph Williamson was the Quaker Master of the Clockmaker's Company in England (Frederick B. Tolles, *James Logan and the Culture of Provincial America*, p. 94.)

[12] This may have been James Wilson (*ca.* 1665–1733), of London. Wilson was noted for his microscopes, and had been commended by John Harris in his *Lexicon Technicum.* See E. G. R. Taylor, *op. cit.*, p. 150.

----

[13] See fn. 11 *infra.*

[14] David Gregory (1661–1708) held for a time the mathematical chair at the University of Edinburgh. He was the first professor to lecture publicly on the Newtonian philosophy. Through the combined influence of Newton and Flamsteed, Gregory was chosen Savilian professor of astronomy at Oxford in 1691, and he became a fellow of the Royal Society in 1692. Gregory's principal work was *Astronomiae Physicae et Geometricae Elementa*, published in 1702. It is said to have been the first textbook based on gravitational principles. Newton thought so well of the book that he had his theory of the moon inserted in it. *Cf.* the biographical sketch in *Dictionary of National Biography* 8: pp. 536–537.

the 3d Vol. of Dr Wallis's[15] works. 'tis a book that I much want, but I should be willing to have all three together.

Finding we are to have a considerable Eclipse of the Sun here in September next, of w[hi]ch you will See nothing, I sett about the calculation of it by various Tables, especially the Caroline[16] & J fflamsteds, as published by Whiston in his Astronomical praelections,[17] which differ Somewhat from those in Sr Jonas Moore's Cursus.[18] The Caroline I find makes the Moons place differ from that found by J fflamsteds above 14'. The latter being so much more forward than the other. I also wrought the Moons place by Newton's Theory & found it above 10' before the Caroline & above 3' Short of fflamsteds. This was the first time I ever attempted to calculate by that new theory, and I expected great helps from Whistons explanations but was much disappointed, in Short they are of no Value. And on this occasion give me leave to observe to thee the great Injustice Students may doe themselves by forming a previous notion that Such and Such a point in the Subject before them is very intricate & difficult. I never in my life calculated the Moons place before this Winter. I first tried by J fflamsteads tables and succeeded without any Stop or difficulty. I then took the theory in hand w[hi]ch for want of tables is more tedious, but is of it Self very plain. I made one attempt and Succeeded, but doubting I had not been exact enough, resolved to lay aside all that work and begin intirely de novo, accordingly I did a few nights after, & before I moved from my Seat, in 4 hours time, went quite thorow it, and I doubt not but if though hast not done it already, thou may easily doe the Same, but perhaps by this time, you may have Tables published, if not, in case it may be of any Service to thee, & thou desires it I shall Send thee some plain directions that may facilitate the work. In the meantime lett me hint to thee, that if thou uses Whiston's help thou wilt find an error in the last numbers that he gives for the first Equations of the M [?] Motion, Apogee & Nodes place, w[hi]ch doubtless thou wilt thy Self easily discover.

But I am Surpriz'd at one thing w[hi]ch is well worth the inquiry into viz why fflamsteed & Newton so widely differ in the method of finding the excentricity.

That they Should differ a little in the quantity of the Excentricity is not Strange but after they have both fixed on their greatest & least, that they should not agree in the intermediate Excentricities I confess puzzles me. The case is thus in Newtons figure (as in the Margin) making BFd double the Annual Argum[en]t, he gives TF for the Excentricity belonging to that Argum[en]t whereas J fflamsted (as I find by trying his Tables) gives TE for it w[hi]ch is a very considerable Difference. I wish thou wouldst satisfy thy self in this point, and from thence thou may also Satisfy me.

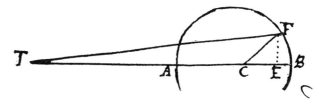

I Shall also further observe on this head, that in finding the Principal Aquation [Equation?] w[hi]ch is that of the Moons place in her Orbit. D Gregory Sayes that Dr Wards[19] method (thou wilt find it in Gregory Lib. 3 Prop. 6 & in Street) is Sufficient but J fflamsted would not (I believe) Say So. I have tried his (J ff's) table by this method & find the difference arise to 2' 40'' in Some places but then applying Bullialdus's[20] Correction [in the next Proposition viz. Prop. 7 & also in Street] it brings it very near to the table & differs at most but about 20''. I wish I could know whether J Flamsted uses Kepler's Method, or this, or what other of his own for this Equation, as it is of the greatest Importance so it is the most difficult of all others. I fancy that Great Man in his Tables, some of them at least, has compound[ed] some other inequality besides that w[hi]ch the title of the Table would import, for Instance, I can by no means bring out the Aquation [Equation?] of the Apogee from His Excentricities to be the Same that He gives. I wish I could ask him the reason. If in health pray inquire of Him about these things & give my very hearty respects to Him.

On comparing the new Theory with that first found by that wonderful Youth Jas: Horrox [Horrocks][21]

---

[15] The reference is to John Wallis, *Opera mathematica* (3 v., Oxford, Sheldonian Theater, 1693–1699). Dr. John Wallis (1616–1703) was a mathematician and a Savilian professor at Oxford University. His collected mathematical works were published, with a dedication to William III, in three folio volumes as cited above. The third volume contains a statement by John Flamsteed "regarding an ostensible parallax for the polestar . . . ." *Dictionary of National Biography* 20: p. 601.)

[16] *Cf.* fn. 40 *infra.*

[17] *Cf.* fn. 41 *infra.*

[18] This citation is not clear, but it must be a reference to Sir Jonas Moore's *A New Systeme of the Mathematicks.* See fn. 42 *infra.*

[19] Seth Ward (1617–1689) was successively bishop of Exeter and Salisbury, but in the middle of the seventeenth century he served for a time as a Savilian professor of astronomy at Oxford. Ward is author of *In Ismaelis Bullialdi Astronomiae Philolaicae Fundamenta Inquisitio Brevis* (Oxford, 1653), a critique of Ismael Boulliau's *Astronomica Philolaica* (1645). In this work Ward advanced a theory of planetary motion simpler and more accurate than Boulliau's. (*Cf. Dictionary of National Biography*, 20: pp. 793–796.)

[20] Ismael Boulliau (1605–1694), a French astronomer.

[21] Logan must have meant Jeremiah Horrocks, who was indeed a wonderful youth. Horrocks (1617?–1641) was an English astronomer who made significant contributions to the science before his premature death at about the age of twenty three. Most of his literary remains were lost during the English Civil

for whose name I have the greatest Reverence, I cannot but admire at their harmony. But I must now conclude after I have told thee that I must send thee back my watch to see if another Spring will make it goe but as well as the most ordinary that are made now, it goes most wretchedly, but I shall try it a little longer[.] if thou buyes any thing for me J Askew will pay thee the money on shewing him this. Pray write by the first Opportunities either directly hither or via N York from whence we have a weekly Post.

LOGAN TO ROBERT HUNTER, PHILADELPHIA, LAST OF JULY, 1718; LETTER BOOKS OF JAMES LOGAN, Vol. II, pp. 180–181, HSP.

The Inclosed from the Gov[erno]r tells my Story about the Quadrant. I heard but accidentally of its going to N[ew] York, just as it was to be sent away, and having never had the use of it[.] Since I knew not when or whether ever it might be here again I was very desirous to make some use of it while in the place & ventured to take upon me I hope not too tardily, to obtain thy leave for it. I gott it but on Saturday last Since w[hi]ch the Skye has been so generally overcast that I have been able hitherto to make but very little use of it.

LOGAN TO ROBERT HUNTER, PHILADELPHIA, SEPTEMBER 2, 1718, LETTER BOOKS OF JAMES LOGAN, Vol. II, p. 182, HSP.

Having ever since I had any acquaintance w[i]th Dioptricks & had occasion to think of Instrum[en]ts suspected that Lights w[i]th glasses might be Subject to great Errors—I was fully convinced of it in the little time I had the use of our Gov[erno]rs Quadrant, w[hi]ch by this time I Suppose may be with you.

I was told indeed that it was proved by Some great Artists in England and reduced to the utmost exactness, yet Dr. Colden who had it in keeping many months, judged it erred a minute or two, but discovering in this what I had before examined, I was mischievous enough to lett it goe home in Such a condition as would convince the Doct[o]r that it is liable to far greater Alterations, at first he might Suppose it has been hurt in the Carriage, but the fault lies only in the object glass of the moving Telescope, w[hi]ch being

so loose that it will turn round w[i]th hard wiping it produces a variation of at least 50 Min:

Tho' this discovery might prove perplexing at first, yet I thought there might be a real Service in it to you both, to let you see by certain Experiences what some have not been very forward to believe, the method of curing it is to take off the top of that moving telescope, which is fastned by two little screws that are easily undone, and between ones finger and thumb putt in at the two ends of that top, when off, to turn the glass so far round till by trials an object on the earth shall give the Same Altitude or depression exactly on or near both the Edges of the Quadrant, for to try that Sight by moving the telescope thus to both edges is the true way of proving & rectifying it, & not by comparing the fixed & moving telescopes with each other.

The fault of the Glass is that the Vertex is not in the middle of it but one edge of it is thicker than its opposite Side, by w[hi]ch means the refracted rayes meet not at the focus in the center, as they ought, nor is it to be mended by new glass better ground, yet after all the Skill & care that can be used in them they are never to be warranted but from experiment[;] they may by trials be brought to give the altitude exact, but then there is the Same Error in the Azimuth w[hi]ch in that Instrum[en]t is of very little Importance. When you have reduced it to exactness the glass ought to be fastned with some little bits of hard wax the Maker having not soldered the Ring close enough down as he ought to have done, but then if he had made a mistake and not Sett it right, it had been incurable.

In dewy nights when elevated to high stars, it must be sometimes wiped as well as from dust at other times, and then it will require to be done tenderly.

Were I to use that Quadrant for any considerable time, I would take it off that Pedestal & fix it in a solid Weighty p[ar]t of Timber whose upper face should be a level[,] be sett exactly horizontal & have a conveniency to sett a pott of water on for the plummet to play in, w[hi]ch w[oul]d sooner bring it to hang steady without motion.

Yet at all times it is convenient to take down on Paper the degrees & minutes that the plummet cutts, as well as those cutt by the fiducial edge of the moving sight, this being done every 3, 4, or 5 minutes the least difference between these two will give the Sun's or Stars greatest Altitude by taking the complem[en]t.

But for a greater certainty I take this to be a good method, viz when the Star is very near the meridian, w[hi]ch may be exactly enough known by its Right Ascension, and a well rectify'd Clock. The Quadrant may be raised or lowred five, ten or twenty degrees where the height of the object will allow it, that is if the plumet at the last observation by its thread cutt 70 degrees or thereabouts it may be sett to 75, 80, 85 on one hand or 65 or 60 &c. on the other hand and after those Changes the Observations should be re-

---

War, but some were rescued. Among them was *Venus in Sole visa* which was published by Hevelius in 1662 as an appendix to his own *Mercurius in Sole.* This publication attracted the attention of the Royal Society, and Horrocks's writings, which were in the possession of Dr. John Worthington, were turned over to Dr. John Wallis for publication. A quarto volume, entitled *Jeremiah Horroccii Angli Opera Posthuma,* appeared at London in 1672. The book includes, among other things, Horrocks' "Theory of the Moon." Horrocks "first ascribed to the moon an elliptic orbit of which the earth occupied one focus, adding a variation of the eccentricity, and a revolution of the line of apsides." Newton, in the third edition of his *Principia Mathematica,* acknowledged his obligations to Horrocks. (*Dictionary of National Biography,* 9: pp. 1267–1269.)

peated as quick as possible, that will help to discover whether there is any error in the graduations, and if I mistake not it will appear that the upper part of the fiducial edge is a little too much filed away w[hi]ch may sometimes produce an error of a minute or two.

Observations by the Sun are most common, but I cannot believe these are to be relied on for exactness, for when near the Tropicks the uncertainty of his greatest Declination, I mean, that of the Eclyptick render them doubtful. Copernicus made it 23°28′30″, all the largest Tables I have of the Primum Moble [Mobile], viz Argols[22] & Vitalis's,[23] are computed to 23°32′ Ricciolus[24] & Hevelius[25] make it, 23°30′20″, the French Academy 23.29′, and fflamstead[26] the Same, tho' 23.30′ is now commonly used, and whether Cassini[27] and and [sic] fflamstead are to be preferred

to Riccioli his Countreyman and fellow labourer, and Hevelius than whom never any man observed more, nor as some have Said with better Instrum[en]ts, is Somewhat doubtful.[28] Again if the observations be near the Equinox there is no depending on any Calculations of the Suns place nearer than 2 or 3 Minutes in Space, and two Minutes and a half in Longitude, near the Equator make one m′ Declination.

What I have said of the Suns place may p[er]haps appear strange, but once on a certain occasion tried it by a dozen several tables, all of them in good repute and none older than the Rudolphine[29] w[hi]ch being built by the great Kepler on Tycho's and his own observations, were long the Standard for Calculation and even these differed from fflamsteads no less than between 8 or 9 Minutes, w[hi]ch makes between 3 & 4 Mins. difference in the declination but besides this one cannot be so exact in the Sun, as in a Star within a Minute or two in the observation.

Again the tables of the Stars places vary so much, that Scarce any can be relied on, Tycho corrected Ptolemy's, Hevelius above all others at that time by observations for above 40 years, undertook to give new and more exact ones, fflamstead has since given others w[hi]ch tis probable may be the best but when one sees how they disagree, we must conclude there is no certainty[.]

The only sure method of finding the Latitude, provided the Instrum[en]t may be depended on, or if we certainly know its Variation, w[hi]ch may be allowed for, seems to be by the Pole Star, by taking its greatest Altitude above the Pole and its least under it. About the 20th of December it may be taken both above and below, the Same night viz. about 6 in the evening and 6 in the morning, if the Skie favours, but it will doe very well to take repeated observations before that time in the evening and others again in the Morning after that day, and the half of these two added together will be the true Latitude. The Same may be done by Some other Stars not far from the Pole, but then the Allowance to be made for the Refraction of the Air, may again occasion an uncertainty, for in the business of Refractions, Authors differ very much, Cassini's tables making it greater than any others (I think) ever did before him.

The Pole Star is the easiest of all others to observe, because of the Slowness of its Motion, for about the

[22] Andre Argoli (1570–1650), professor of mathematics at the University of Padua, was the author of *Tabulae primi mobilis* (Rome, 1610) and *Secundorum mobilium Tabulae* (Padua,1634). (See *Nouvelle Biographie Générale* [Paris, 1891] **3**: p. 134.)

[23] Hieronymus Vitalis, a Theatine of Capua, was the author of *Lexicon mathematicum astronomicum geometricum hoc est rerum omnium ad utramque immo et ad omnem fere mathesim quomodocumque spectantium collectio et explicatio* (Paris, 1688). This work contained much astrological information. (Lynn Thorndike, *A History of Magic and Experimental Science* [8 v., New York, Columbia University Press, 1958], **8**: p. 329.)

[24] Jean-Baptiste Riccioli (1598–1671), an Italian astronomer, was particularly concerned with the correction of what he believed to be errors in astronomy, chronology, and geography. Critical of Copernicus and Galileo, Riccioli wrote a number of scientific works. The most important of them was *Almagestum novum, astronomiam veterem novamque complectens* (2 v., Bologne, 1651). (See *Nouvelle Biographie Générale* [Paris, 1891], **3**: p. 134.)

[25] Johann Hevel (1611–1687), called Hevelius, had his own observatory at Danzig at which he employed telescopes of greater power and larger focal distances than those previously used. He made an accurate study of the moon's surface, and described his astronomical observations in books which Wolf calls "the finest products of seventeenth-century descriptive astronomy." (A. Wolf, *A History of Science, Technology, and Philosophy in the 16th and 17th Centuries* [New York, Macmillan, 1935], p. 182; Giorgio Abetti, *The History of Astronomy*, translated by Betty Burr Abetti [New York, Abelard-Schuman, 1952], p. 125. Also see Lynn Thorndike, *A History of Magic and Experimental Science*, **8**: p. 336.) According to *Nouvelle Biographie Générale*, **25**: p. 286, Hevelius's real name was Hovel, or its diminutive Hövelke.

[26] John Flamsteed (1646–1719) was the first director of the Royal Observatory established in 1675 at Greenwich Park. His accumulated stellar observations were published in 1712 as *Historia Coelestis Britannica*. Revised and reedited in 1725, this catalog includes 2,866 stars. (Gerard de Vaucouleurs, *Discovery of the Universe: An Outline of the History of Astronomy from the Origins to 1956* [New York, Macmillan, 1957], pp. 62–63.) See fn. 30 *infra*.

[27] Gian (or Giovanni) Domenico Cassini (1625–1712) was the first of four generations of Italian astronomers who bore the name of Cassini. He taught for a time at the University of Bologna. Then upon the invitation of Louis XIV he became director of the Paris observatory in 1669. His calculations of the earth's distance from the sun, the distance of Mars from the Earth, and the dimensions of the planetary orbits were more accurate than those of earlier astronomers. (Giorgio Abetti, *The History of*

Astronomy, pp. 127–131; Lynn Thorndike, *A History of Magic and Experimental Science*, **8**: p. 335.)

[28] It is now generally accepted that the plane of the ecliptic is inclined to the celestial equator by about 23 1/2°.

[29] Tycho Brahe (1546–1601), the Danish astronomer, began the astronomical tables which were completed by Johannes Kepler (1571–1630), under the patronage of Rudolph II of Germany. The *Tabulae Rudolphinae* were published in 1627. (John Allyne Gade, *The Life and Times of Tycho Brahe* [Princeton, Princeton University Press, 1947], pp. 190–191; Carola Baumgardt, *Johanne Kepler: Life and Letters* [London, Victor Gollancz, 1952], pp. 170–171.)

Meridian it varies one minute in altitude in no less time than about 25 min. but its declination at this time is somewhat uncertain, for tho' Astronomers have been very careful to take it, and its distance from the Pole has often been exactly taken, yet as that w[i]th all others has a yearly motion in Longitude, its nearness to the Pole makes it difficult to find its R[ight] Ascension w[i]th certainty and on this depends the fixing of its Longitude, from whence its Latitude in process of time its declination is alwayes to be calculated de novo.

I thought of ending this Lett[e]r short of the Sheet, but must now I perceive close it on another, therefore I shall proceed a little further.

That you may see how unfixed the true place of that Star is to this day, I shall annex its Longit[ude] & Latit[ude] according to several Authors as Hevelius gives them in his Catalogue, all computed to the end of the year 1660, by w[hi]ch their Variations will appear in both.

Our Gov[erno]r procured a manuscript Copy of fflamsteads Tables of the Stars of the Zodiaque, and 3 other Constellations, w[hi]ch I have now by me, and am transcribing them into my Hevelius, but in these the Pole Star is not, nor any other of the Northern Constellations, the Printed Book is not to be had for Money, having never been published, and since the Kings accession, the Author having gott them into his possession, burnt them as we are told, because published or rather printed without his leave by the Royal Society with some Alterations, of w[hi]ch I have heard him grievously complain.[30]

Doct[o]r Colden has in the 2d Vol of Harris's Lexicon[31] the places of above 50 of the brightest stars, computed to the year 1710, and there one would expect exactness, being so modern, but they are wholly taken from Hevelius. The Longitude is there given by a[d]ding 42'15" for 50 years to their places in Hevelius tables calculated for the end of the year 1660, and the R Ascensions and Declinations are also taken from the same Author who has a particular table to show their Variations in 100 years. So that besides the Errors in the print of which that book is full, these are not to be depended on.

J. Alexander tells me he has Halley's Hemispheres,[32] w[hi]ch if they are the same that are generally called so & w[hi]ch I have, are furnished w[i]th the RA & Decl[ination] of 80 of the principal Stars calculated to the year 1678, but these are not Alleys [Halley's], I believe, but were only calculated, as I judge, by Seller[33] from Tycho's or Ricciolus's, w[hi]ch are much the sam[e]. I have tried some of them & do not find them to agree w[i]th fflamstead in such Stars as are in the ms Catalogue I have mentioned.

By the next Post after this, if I can find time, I may Send thee the R[ight] A[scension] & Decl[ination] of about a score of those in Harris's tables calculated to this time from fflamstead's Longit[ude] & Latit[ude] w[hi]ch you may compare with the other, but I shall principally choose those that can be observed from this time till next Spring.

I have now been very tedious and p[er]haps am too officious, but what I write I hope will be taken with the same goodness as usual,—what I have said of the Quadrant is for thy particular Information, for the Doct[o]r wants no part of it, & what I have added concerning the Stars is because I doubt you are not so well furnished w[i]th Books of that kind as I happen to be at this time. I am with great respect

Thine to Command in all things

J. L.

The place of the Pole Star 31°10ᵐ 1660 according to these following Authors from Hevelius's Prodromus at Dantzick 1690[34]

	Hevelius	Tycho	Princ. Hessia	Ricciolus	Ulug Boig	Ptolemy
Longit.	23°51'22"II	23.53.30"	23.45.7"	23.53.0"	23.29.0"	21.45.0
Latit.	66.3.0	66.2.0	66.1.15	65.59.50	66.27.0	66.0

These places being reduced to the beginning of the year 1720 by adding 49'30" to their Longitude for 59

---

[30] After Sir Isaac Newton became president of the Royal Society there was a long dispute between Flamsteed, on the one hand, and Newton and Edmund Halley, on the other, over the publication of Flamsteed's astronomical observations. The *Historia Coelestis* was printed toward the end of 1712, after it had been considerably edited by Newton and Halley who believed that Flamsteed's observations were full of errors, and Flamsteed strongly objected to the publication. There was a change in Flamsteed's favor after the death of Queen Anne in 1714. The remaining 300 of the 400 printed copies were procured for Flamsteed by the order of Sir Robert Walpole, and Flamsteed burned them up. The *Historia Coelestis* was therefore a very rare book after 1714. (*Dictionary of National Biography*, 7: pp. 245–246.)

[31] John Harris, *Lexicon Technicum; or an Universal English Dictionary of Arts and Sciences, explaining not only the terms of Art, but the Arts themselves* (2nd ed., 2 v., London, 1708–1710). The first edition, in one volume, appeared in London in 1704. John Harris, D. D. (1666?–1719) was the author of a number of books on science and theology. (*Dictionary of National Biography*, 9: p. 14.)

[32] Alexander must have referred to plates showing the celestial hemispheres which Halley had made and which John Seller had published. According to Edwin Wolf 2nd, these plates are not in the Library Company of Philadelphia, but a later edition of them appeared in Halley's *Atlas Maritimus* (London, 1728). Wolf does not "doubt that Logan owned a good many separate maps and charts which were not preserved in the Library. I assume that the two star maps were among them." (Edwin Wolf 2d to the editor, Philadelphia, April 22, 1968.)

[33] John Seller, who flourished about 1700, was a publisher of maps, charts, and geographical books. (See *Dictionary of National Biography*, 17: pp. 1165–1166.)

[34] *Prodromus Astronomiae exhibens fundamenta* (Danzig, 1690).

years at 50′20″ for one years motion, I have calculated their Declination as also their R[ight] Ascension, I believe with some exactness, as follows:

Compl[ete] the Declination of the Pole Star 1 Jan[uar]y 1719/20 from the foregoing Longit[ude] increased

	Hevel.	Tycho	Pr. Hessia
Distance from the Pole	2°11′30″	2°10′47″	2.14.14 1/3
R[igh]t Ascensions	9.51.30	10.28.24	9.38.24

	Ricciolus	Ul. Boig	Ptol.
	2.11.30″	2.16.48″	3.1.54″
	11.12.23	1.31.16	6.21.53

But these Ascensions will be of little Service to you. the Declinations may, for you may pitch on such a Medium as you may judge nearest.

Hevelius was consul of Dantzick & w[i]th vast and chargeable Instrum[en]ts observed for about 50 years encouraged to it by the King of Poland & divers other Princes[,] but he used no glasses.

Wm Landgrave of Hesse observed at Cassel w[i]th very good Instrum[en]ts and Able Artists about the same time.[35] Tycho did or rather before him he took the places of between 4 & 500 Stars from the first to the 4th Magnit[ude]. his Observations Agree as well with fflamsteads as most of the rest but in this Star he varies much from Tycho & Hevelius.

Ricciolus of Bolognia tho' he pretended to give a more exact Catalogue of the whole observed himself but about 100, he altered Tycho's Tables[.]

Ulug Boig was nephew to Tamerlane & observed about the year 1430 at Samarcand to the eastw[ar]d of the Caspian Sea, his Tables were published by Doct[o]r Hyde of Oxford from three Copies in that Library A D 1665 in Arabick and Latin being translated by the Editor.[36]

Hevelius published all these with his own reduced to the y[ea]r 1660[.]

I run over the Inclosed two or three dayes agoe expecting that I must have this day been at Newcastle with our Gov[erno]r who is now there. I intend it wholly for thy own View only, and therefore if there be occasion to communicate to Doct[o]r Colden these Calculations of the Pole Star R[ight] A[scension] & Decl[ination] it may be done separate from the rest.

LOGAN TO HUNTER, PHILADELPHIA, 17 SEPTEMBER 1718, LETTER BOOKS OF JAMES LOGAN, Vol. II, p. 186, HSP.

I am just now favoured w[i]th thine, and am obliged to answer it immediately if at all by this Post, and therefore request to be excused if not so full on those heads as otherwise I might be.

I hope J[ames] Alexander has been for some dayes sufficiently recovered to be w[i]th you there, he was acquainted here if I mistake not with all the varieties of the Quadrant that I mentioned in my last.

· · · · · · · · · · · · · · · · · ·

During the time I had the Quadr[an]t we had very fine clear dayes or nights, and my observations by the Sun disagreed so much w[i]th those I made by the Stars especially the Northern ones that I could fix on nothing certain for our Latitude. Dr. Colden gave our Gov[erno]r a Note of his observation mentioning particularly one made in the Gov[erno]rs p[re]sence by Lucida Aquilae whose declination he made 8°9′ but it should be 10′ for so much it is now if in 1710 according to his table it was as he took it viz. 8°9′ he rec[k]oned our Latit[ude] 39°57′ and I believe this is not much above a minute out of the way.

By several Stars after the Quadrant was rectify'd especially those of Sagittary [Sagittarius] which now pass the Meridian before night I made our Latit[ude] 39°58′30″ or 59′, but calculating the declination of those stars from fflamsteads places I found them vary considerably from Hevelius's and Tycho's. And it is to be observed generally that fflamstead gives the Latitude of most of the Stars in those tables of his w[hi]ch I have more Southerly than the other two Authors, w[hi]ch will alwayes make the Latitude of the place less-but this being most remarkable in the most southerly Stars, w[hi]ch have the least Meridian Altitude, (for in the higher ones he disagrees less) I imagine he makes the Refraction greater, if we had his table of Refractions it would clear that point, but what Strengthens me in this opinion is that Cassinis or the French Academies Tables[37] after Kepler[,] Hevelius and others had very considerably lessen'd the allowance made for Refractions, have very much increased it, and as fflamstead and Cassini generally agreed in other points so tis probable they might in this.

---

[35] William IV, Landgrave of Hesse-Cassel (1531–1592), was an avid student of mathematics and astronomy. The results of his astronomical researches were published as *Coeli et Siderum in eo errantium Observationes Hassiacae* (Leyden, 1628). (See *Nouvelle Biographie Générale*, 22: pp. 638–639.) Logan elsewhere referred to William as the Prince of Hesse.

[36] Ulug Begh (or Beigh or Boig) (d. 1449) was the grandson of Tamerlane and an important Oriental philosopher in his time. At his observatory in Samarkand, built in 1420, he prepared new planetary tables and a new star catalog, the first since Ptolemy's. (J. L. E. Dreyer, *A History of Astronomy from Thales to Kepler* [2nd rev. ed., Dover Publications, 1953], p. 248.) Ulug Boig's catalog of fixed stars was included in Thomas Hyde (1636–1703), *Tabulae Longitudinum et Latitudinum Stellarum fixarum ex observatione principis Ulugh-Beighi; accesserunt Mohammed Tizini Tabulae Declinationum et rectarum Ascensionum, arab. et lat., cum commentariis* (Oxford, 1665) with notes by Hyde relating to the names of the stars. (*Nouvelle Biographie Générale*, 25: p. 694.)

[37] The Paris observatory, directed by Cassini, was built for the Académie des Sciences in 1672. Cassini's astronomical tables were the property of the Académie des Sciences. (A. Wolf, *A History of Science, Technology, and Philosophy in the 16th and 17th Centuries*, p. 63.)

Therefore the Declination of the most eminent Stars in Capricorn Aquarius Cetus and Aries (w[hi]ch are first in Ord[e]r that came on our Southern Meridian from this time for 3 Months to come) calculated from fflamsteads places to the year 1720 Commencing which is now but about 15 months before us. I give you also fflamsteads Longit[ude] and Latit[ude] as they stand in his tables computed to the year 1690. Commencing to w[hi]ch places I added just 25 Min[utes] of Longit[ude] for the 30 y[ea]rs and from thence calculated their Declination[.] I had not time to doe the right Ascensions but if these Declinations be right the others may easily be done by one single Operation from them whereas these are more troublesome. I confess might have lett this alone every Artist desiring himself to be Satisfied of the work but I was willing to doe it for my own use and being furnished with better Tables of Signs &c Sharp and commonly to be mett w[i]th if my work be true w[hi]ch I cannot as yet fully answer for, it will Save a great deal of trouble, for in solving oblique triangles that require two operations the 4th and 5th Arches w[hi]ch are found must be taken exactly[;] otherwise they may occasion a considerable variation in the solution. I wrought these in those Arches to the tenth part of one Second w[hi]ch by common tables would be much more tedious than by mine. But I have not time to add any thing more at present hearing the Post is just going. Shall only run over below what I can of what I have mentioned & must here close.

	Declin[ation]	
Sagittaries left Shoulder	26° 3' 50'' So	
Capricorn North horn	13.22.58	S
South horn	15.38.6	S
foremost in the tail	17.44.33	S
The following in the tail	17.22.8	S
Aquarius left Shoulder	6.47.42	S
right	1.39.59	S
in his right arm	2.47.26	S
ffomahant[?][38]	31. 5.14	S
Scheat Aqurii[?]	17.18.13	S
Knot in the Line of Pisces	1.23.34	N
first Star of Aries	17.54. 3	N
second Star of Aries	19.25. 8	N
Th[e] bright Star of Aries	22. 7. 5	N
Mandibula in the Whole of his jaw	2.57.49	N
Lucida Caudeae[?][39]	19.31.48	S

I shall add more by next with the Long[itude] & Latit[ude] I mentioned.

LOGAN TO HUNTER, PHILADELPHIA, SEPTEMBER 25, 1718, LETTER BOOKS OF JAMES LOGAN, Vol. II, p. 188, HSP.

---

[38] This may be Fomalhaut, a star of the first magnitude in the constellation Pisces Australis. It was called Fomahant in John Harris's *Lexicon Technicum*.

[39] This may mean "brightest star in the tail."

I return my hearty thanks for the News w[hi]ch is very great but we hear further by some other way that the Spaniard has made a descent on Sicily & that K[ing] Philip has been proclaimed at Palermo Sovereign of the Island, If so Savoy who has so long been too hard for the rest of Mankind may himself become a Dupe, if not powerfully assisted by those who p[er]haps may not think themselves so much in his Debt.

Since my last I have not had leisure to doe anything further in Calculation, but I could now Subject even fflamsteads Observations almost as much as the rest my reasons for which I in some part hinted in my last, and have since this further Strength added to them.

Last Week Jupiter pass'd Regulus or the Lions heart, & very near to it, for 2 or 3 Mornings before I observed their distance and found it decrease from 40 Min[utes] to 17, nor were they at this last observation come to a partile Conjunction[.] The next morning when they were to be so the sky was overcast, but I really think they could not then be above 15 or 16 Min[ute]s distant, by fflamsteads Latitude of Regulus and the Caroline tables[40] for the Motions of Jupiter they must have been 20 Min[utes] Assunder when nearest. The Obliquity of Planets Orb has been so very often tried that we cannot Suspect an Error of 5 Min[utes] in it especially when not very far from his Node, tis most probable therefore that those who make the Stars Latitude the greatest (for it is North & Jupiter was to the Northw[ar]d of the Star) must be most in the right and this Hevelius does by above 2 Min[utes] more than fflamstead and the Prince of Hesse by almost 3 but Tycho and fflamstead in this happen nearly to agree. My method of observing I think could not err much, for such observations are very easy to a Radius of 6 or 8 feet by only erecting 2 needles perpendicularly near one end of a Square Stick of that Length and applying the other end to the eye, then measuring both the length of the Stick and the Needles distance by a very exact Scale and computing the Arch from a table of Sines.

If you would be pleased to communicate your greatest Altitudes or Zenith distances of the Stars you observe by, I would compare them w[i]th the Latitude of those Stars according to the Several Tables I have.

---

[40] The reference is to Thomas Streete (or Street), *Astronomia Carolina, with exact and most easy Tables and Rules for the Calculation of Eclipses . . . To which is added a series of observations on the Planets, chiefly of the Moon, made near London . . . in order to find out the lunar theory a posteriori . . . By Dr. E. Halley.* Halley's tables of nearly 200 lunar observations were added as an appendix to the second edition, published in 1710. (*Dictionary of National Biography* 8: p. 989.) Logan had a copy of the 1710 edition, and it is now preserved at the Library Company of Philadelphia. (*Annual Report of the Library Company of Philadelphia for the Year 1961*, p. 44.)

LOGAN TO HUNTER, PHILADELPHIA, OCTOBER 2, 1718,
    LETTER BOOKS OF JAMES LOGAN, Vol. II, p. 188,
    HSP.

Late last night I was favoured with thine inclosing 4 Observations. I examined the Declinations and found the Latitude to arise from them all between 40.37' & 40.40—when due allowances are made.

Having omitted last Post to p[er]form my promise I this morning being reminded of it by J. Alexanders kind Let[e]r Sett about drawing out the Longit[ude] & Latitude I mentioned before I happened to have more time than usual but not enough[.] however I think they are exact having carefully compared them. You will choose those that agree best with your carefullest Observations by which you may be able I believe to correct even the best of them.

I chose to give them as they stand in the Several tables without any Reduction. If fflamstead be preferred those Stars in the Whole may easily be calculated as also those in Orion of which I should have added Several more would time have allowed it,— those I Say will be easily calculated from those R[ight] Ascensions & Declin[ation]s he gives himself with the Variation in each for 1 Degree of Longit[ude] but how long time the Stars Spend in moving a degree he does not fix & Hevelius makes the yearly Motion

50"50''' And has Accordingly allowed I suppose to bring up the other Astronomers tables to his own viz 1660[.] But I cannot find our Engl[ish] Astronomers allow any more than 50" and by that I computed those that I did adding 25' for 30 years. Tis according to this allowance that Hevelius's Variation of the R[ight] Asc[ensions] & Decl[inations] for 100 years is calculated.

I mentioned in my last that Authors differ about the Allowance to be made for Refractions. I have now [,] that you may be furnished with what tables I have [,] caused my Lad to draw them out of the Several Writers & putt them into one table which will at least shew the variety of Sentim[en]ts that there is about them[.] It comes inclosed with the other.

I have not forgott that I am in debt as to the translation[.] I thought of these Astron[omical] Subjects only by necessity & not of Choice at this time of year. And to drudge at a little Calculation is but toil only without requiring what the other indispensably craves. I have not therefore been able to think of it to purpose but as soon as Capt[ain] Annis Sails I hope for some Leisure. The speedy Return of the Post always Straitens us as now it does particularly.

I shall here add some of my last nights work for your Latit[ude] & request more of your Observations[.]

Lucida Aquilae Alt.	57.32			
Decl. Bor.	8.10			
	49.22			
Poles Elevation or Latit.	40.38			
Humer Sen^n	42.35		42.35	
Deliin[?] Aust. Hevel.	6.44.35	fflamstead	6.47.42	
			49.22.42	
	49.19.35		40.37.18	
Latit. or P Elev.	40.40.25			
O Pegasi alt.	58.0.0			
Decl. Hevelii Bor.	8.39.26			
	49.20.34			
Latit.	40.39.26			

ffomahant Alt.	–18.15.0	—18.15.0	
Decl. Heveli	–31. 0.38	—31. 5.14	but this should be corrected by adding the Refractions to the Latit.
	49.15.38	49.20.14	
Latit.	40.44.42	40.39.46	

7br 27th O's place at York at noon by fflamst–15.1.28 Decl. 5°55'42"

but by the Caroline Tables—   14.57.8 Decl. 5-54-4

O Alt.	43.44.15	Altit. O's Center Corr.	43:27:6–43:27:6	
Ded. Refr.	1.1	Decl. fflamstead Tabl.	5.55.42 Caroline	
& O Sendiam	16.8		5.54.4	
	17.9		49.22.48	49.21.10
Altit. O's Center	43.27.6	Latit.	40.37.12	40.38.50

It is also to be rememb[e]red that Hev el[ius] & Tycho make the obliquity of the Ecliptick above 23.30 but fflamstead only 23°29'[.]

LOGAN TO HUNTER, PHILADELPHIA, OCTOBER 7, 1718, LETTER BOOKS, OF JAMES LOGAN, Vol. II, p. 190, HSP.

Happening to have more leisure last 5th or Thursday forenoon than usual I sett about discharging a former engagement in drawing out the Longitudes & Latitudes of Such principal Stars as would be most Suitable for your observation for some months but I endeavoured to doe it so fully that it cast me a little behind hand and when I sent my letter to the office just as it comes inclosed save that I have since opened it the Post had been gone about a quarter of an hour. The Letter itself will show it was done in very much haste. Therefore pray be pleased to destroy it all except the Calculations and I shall here repeat what I judge to be worth notice in it Somewhat more distinctly.

The table I send gives the Longit[ude] and Lati[tude] of the Several Stars according to Hevelius, Tycho, and the Prince of Hesse who have been the best observers we know of fforeigners. For the Pole Star I formerly added likewise Riccioli's[,] Ulug Boigh's and Ptolemy's but the first of these observed but few himself, yet from these few undertook to correct Tycho and therefore is not to be depended on [.] The two latter are generally too wide from the rest to trouble you w[i]th them.

These are all as the Table shows reduced to the year 1660 compleat by Hevelius from whom I only copy them in the next column follow[ing?] their R[ight] Asc[ension] & Declin[atio]n to the same time calculated by Hev[elius] to his own places & next the difference of these for one hundred years from that time showing by the words or Letters ad[d]ed to them when the Declin[ation] is to be added, and when subtracted w[hi]ch is of great use because that Column will be equally Serviceable for any other Author's calculation of the same stars, and will save a great deal of trouble.

After the double line follow fflamsteads Longitt[ude] and Lattit[ude] of the same reduced to the beginning of the year 1690, as he gives them himself[.] These I have from the Papers I formerly mentioned to be brought over by our Gov[erno]r but as they have only the Constellacon of the Zodiaque with Cetus Orion and Serpentarius I could not give these of Pegasus from him.

In these three last viz. Cetus &c. the Table is much more full and Compleat than in the rest for it gives the R[ight] A[scension] & Decl[ination] to the same time & also the Variation of these for one degree of Longit[ude], but he takes not upon him to determine how long the Stars are in moving forward that degree, or more properly speaking what the praecession of the Equinox is. Hevelius makes it 50″, 52‴ and has I suppose accordingly computed his places & those of the other Authors, but I find the common Allowance made by our English astronomers, and I suppose the french also is 50″ in a year, and by this I reduced those few whose R[ight] A[scension]s & Declin[ation] I give in the last column calculated from fflamsteads Lo[n]gitt[ude] & Latit[ude] to the beginning of the year 1620.

Besides this difference it is also to be remembred that all the rest (tho' I am not sure of the Pr[ince] of Hesse who perhaps might follow Copernicus) make the Declin[ation] of the Ecliptick to be above 23.30 but fflamstead only 23.29 as also does the ffr[ench] Academy.

Because there is a great regard to be had to the Refraction of the Air I caused to be drawn out into one, several tables of it, I have not fflamstead's & the Academy's only for the Sun, not the Stars.

To the end of the Tables of Places I have added the Longit[ude] & Latit[ude] of the Cranes head & that in the Phoenix's Neck both to be seen in the South about this time very low. These places are from Halley's observations of them in St. Helena where their Merid[ian] Altit[ude] is between 6 & 70 degrees & therefore clear from refraction. Now supposing Halley's to be exact, observations of the same Stars here where the height of one of them is but about 10 degr[ees] & the other 5 or 6 will give a good proof of the Refraction at those Altitudes[.] I therefore should be very glad to see a few careful observations of them, for they would afford me at least a considerable Light, the season for observing the Crane will be over in a little time, but the Phoneix continues longest being above 2 hours later[.]

Another method I should take (if I had an Instrum[en]t to try the Refractions of this Climate would be to observe all the Stars not less than the 3d Magn[itude] and not above 36 deg[rees] or thereabouts from the Pole as they come on the Meridian either above or below Successively[.] Such are the Stars of the 2 Bears the Draggon Cephous Cassiopeia & one or two in the Swans North Wing w[hi]ch never sett in these Latitudes. Such observations might certainly fix the Refractions because the Same Star would be observed at Such very different Altitudes. And I should think this the better worth while because as our Air is much heavier here than in Engl[an]d So p[er]haps the Refractions may differ.

If you want any continuation of this table of the Stars I shall send it with everything else that may be acceptable. . . .

LOGAN TO SIR WILLIAM KEITH, PHILADELPHIA, OCTOBER 9, 1718, LETTER BOOKS OF JAMES LOGAN, Vol. II, p. 193, HSP.

It might be more reasonable in me to defer adding anything further on the Subject of Refractions till I saw whether my Proposals were approved, but what I have now to offer if putt in practice this year will not admit of much delay[.]

I mentioned in my last that went w[i]th the Express two considerable Stars to the Southw[ar]d well worth observation on acco[un]t of the refraction, but I happened since viz. last night and the preceeding when the Sky was not very clear at least not so serene as it is some frosty nights to observe very plain two other Stars in the Crane w[hi]ch are both of the 2d Magni-

tude and they may also be seen with you in a very bright night near a degree above the Horizon.   I hereupon reduced them from Halleys places & tables to the 1st January 1719/20 and calculated their declination as below, from whence their R[ight] Ascensions may easily be computed.   If these with the other two whose declinations I have also calculated be carefully observed for some nights, they will give no inconsiderable Light to Refractions.   If such Observations be made I request they may be committed with others to [me].

	Magn.	Longit. 1720 (Jnews?) Latitud. Hallieu	Latitude	Declination	R. Asc.
Caput Gruis	3	13° 30′ 50″ =	22.55.OA	30° 33′ 28′ ′A	324 : 11.40
In Ala dextra	2	11-55-50 =	32-47-OA	48-31-10A	336.17.00
In Ventre	2	18-16-50 =	35-20-OA	48-18-40A	
Phoenicis Caput.	2	11-26-50 =	40-33-OA	43-50-11A	3-1-50

These may be added to the other Table.   I have since added the R[ight] Ascension[.]

LOGAN TO HUNTER, OCTOBER 16, 1718, LETTER BOOKS OF JAMES LOGAN, Vol. II, p. 194, HSP.

. . . .

I return my hearty thanks for the observations w[hi]ch are made, I see to a Surprizing nicety by an Instrum[en]t divided only to 2 Min[utes] but those two Several observations of the Sun show nevertheless how difficult it is to be exact in taking the height of that vast luminous Body[.]   You differ in your Work 2 Min[utes] or thereabouts, for tho' the Paper sent me mentions fflamsteads Tables in the first & Whistons[41] in the 2d, yet these Tables, if by the first, are meant those in S[i]r Jo. Moore's Systema,[42] differ from each other but about 1′30″ in the Suns place & therefore not above 35″ in the Declination.   So that all the rest of the Difference returns on the Observations, but both these Tables are fflamsteads.

I cannot yet discover that I committed any Error in giving the place of Caput Gruis[.]   In my table for the year 1700 compleat its Long[itude] is 13°15′0″ and its Latit[ude] 22°55′[;] to the first of these I added 15′50″ for 19 y[ea]rs and if I made any mistake in the Calculation of the R[ight] A[scension] &

Decl[ination] it may easily be discovered, but either your Observation or Halleys is enormously out of the way.   Those observations to Settle the Refractions can be of no use till the Latit[ude] of the place is fix'd which I hoped you would have done by this time by the Pole Star, the Slowness of whose Motion renders it by much the most proper of any for that purpose[.] it now comes on the Meridian between 11 & 12, but did not until after 3 when I observed it, yet I sate up two or three nights for it.

Hevelius[,] Tycho[,] and the P[rin]sse of Hesse differ not much in those Stars of the great Bear.   I give their places of those 7 in the Wain below.

I am puzzled how to understand the Error of the fiducial Edge of the Quadr[an]t.   I took notice to J[ames] Al[exander] when here & in my Lett[e]r to thy Self that it was not exact but appeared filed off too much tow[ar]d the upper part of it next the Center.   Your Directions are to add the Allowance for this Error to the Altit[ude] & to Subtract it from the Zeniths distance by w[hi]ch it appears to me that you reckon the upper part of it the truest & make the Allowance for the lower, or else I am at p[re]sent too dull to understand it, if you keep the fix'd tellescope uppermost in observing as I perceive you doe in those where the degrees are mentioned that the Sights & the plumet Cutt.   but 'tis probable I may mistake, or else that you place the error to the lower part, w[hi]ch I am now p[er]swaded you doe & therefore am Sorry I wrote this.   Pray pardon my excessive freedom.   I forgett my Self on these Subjects. . . .

LOGAN TO HUNTER, PHILADELPHIA, JANUARY 6, 1718/19, LETTER BOOKS OF JAMES LOGAN, Vol. II, p. 202, HSP.

. . . .

I am sorry your Observations were so long delayed. I rec[ei]vd some from J[ames] Alexander, but have not time to examine the Latit[ude] from them [;]

[41] William Whiston (1667–1752) succeeded Newton as Lucasian professor in 1703, lectured on mathematics and natural philosophy, and was one of the first to popularize Newtonian theories. He was deprived of his professorship in 1710 because of his unorthodox theological views.   Among Whiston's scientific works are the following titles: *Praelectiones Astronomicae* (1707) and *A New Method of Discovering the Longitude* (1714).

[42] Sir Jonas Moore (1617–1679), an English mathematician, served as surveyor-general of the ordnance during the reign of Charles II.   His principal work, *A New Systeme of the Mathematicks: containing . . . arithmetik . . . practical geometry, trigonometry . . . cosmography . . . navigation . . . the doctrine of the sphere . . . astronomical tables . . . a new geography*, etc. appeared posthumously in 1681.   The astronomical tables in the book were by John Flamsteed.

only I perceive those of the Pole Star make it above 40°42'. I shall try the rest & be thankful (as for these) for more.

. . . .

LOGAN TO HUNTER, PHILADELPHIA, 22d JAN. 1718/9, LETTER BOOKS OF JAMES LOGAN, Vol. II, p. 202, HSP.

Had it not been for this last Storm of Snow I de-sign'd to have waited on thee At Amboy this week but now must defer it till the Roads are more practicable.

I am favoured w[i]th a considerable Number of observations from J[ames] Alexander for w[hi]ch I heartily thank you, but am surprised to find that every one of them make the Latit[ude] above 40°40' when those in the ffall generally made it under. I wish they had tried the Same Star, especially the Pole, as being very Slow in motion w[i]th the Quadr[ant] in various Positions as formerly hinted which would show whether there were not some variation or Error in the graduations of the Instrum[en]t or in the working of the Sights on the pin at the Center w[hi]ch requires such exactness in the workmen that it is difficult (to me at least) to imagin[e] how they can possibly attain to it. But since both Southern & Northern Observations agree in making the Latit[ude] more, 'tis a great Argum[en]t that it lies not in the Instrum[en]t. Something to the same purpose to J[ames] A[lexander] & give him places for few more Stars.

LOGAN TO HUNTER, PHILADELPHIA, JANUARY 22, 1718/19, LETTER BOOKS OF JAMES LOGAN, Vol. II, p. 203, HSP.

I heartily thank thee for the trouble thou hast taken in copying out the Observations for me, but am Sur-prized to find them so generally increase the Latitude beyond what it was believed to be by your former ob-servations in the ffall. Those of the Pole Star, I should think, as I have often said, ought to be more depended on for divers reasons. I perceive thou has wrought none of them but those whose greatest and least Altitudes have been taken, and Rigal in Orion by fflamsteads Declination, but I have tried all those of that Constellation both by his & Hevelius's declin-[ation][.] By Rigels decl[ination] in fflamstead 8°33'36" the Lat[itude] as thou makes it is 40.41.32 by Hevelius decl[ination] 8.31'17" (Subtracting nothing for the Refraction, for in his Declinations we must necessarily use his table of Refractions) it is 40°42'43". Orions left shoulder fflamst[eed] Decl[i-nation] 6.3.48" Latit[ude] (the Refraction deducted) 40.43.9"[.] Hev[elius] decl[ination] 6.41.28" Lati-t[ude] 40.43.8". His right shoulder fflams[teed] Decl[ination] 7.19.3" Refraction from the Galic Tables, Lat[itude] 40.39.52" Hev[elius] decl[ina-tion] 7°20'59"—no Refraction Lat[itude] 40.41.9".

But I admire at Sirius the most of any for Hevelius & Tycho agree in his Longit[ude] to 1" & in his Latit[ude] to 5" & the Pr[ince] of Hesse differs only 45" yet I find the Latit[ude] by Hev[elius] Declin-[atio]n of him is 40°45'42" So that here I doubt there is an Error in the Observation, But Riccioli makes the Latit[ude] of that Star 2' more than Tycho & Hevel[ius] w[hi]ch would bring the Latit[ude] of the place to about 44'.

On the other page I give thee the places of the prin-cipal Stars of lesser Bear & Cassiopea from the several tables. I have not now time to add more.

I wish you had taken the Altitudes of the Same Star the Same night at various positions of the Quadrant to see whether any difference would arise from thence, It may be repeatedly done you know with the Pole Star because of the Slowness of its motion, and twice with any other if you carefully compute the R[ight] A[scension] of the Star and have a well regulated Watch for from thence you may know the time of the Stars being on the Meridian to less than a Minute in time. It might also be adviseable p[er]haps to putt down in your Collection the exact time of the observa-tion of each greatest & least Altit[ude] upon Ac-co[un]t of the R[ight] Ascension.

I hope now thou wilt have more leisure at N[ew] York in thy office than heretofore & doubt not of they endeavours to oblige one in the point requested. I am sorry it must prove so troublesome, but it must be considered accordingly.

LOGAN TO JOSEPH WILLIAMSON, PHILADELPHIA, OCTO-BER 5, 1720, LETTER BOOKS OF JAMES LOGAN, Vol. II, p. 228, HSP.

Our Gov[erno]r was of opinion that his Quadrant was perfectly exact having gott his fr[ien]d Pound[43] to examine it most accurately, but having had Occa-sion to try it in Settling the Bounds between the Provinces of New York and New Jersey by the Lati-tude w[hi]ch required the utmost Nicety We found not only a Diff[e]r[ence] between the two ends of it that is using it direct or Inverted, for which it was easy to make due allowances, but it was found that it would differ 3 or 4 nay 5 minutes by taking the Angle be-tween the Plummet & fiducial edge of the moving telescope that is the Zenith distance at Several places of the Limb w[hi]ch is Error. I think never to be mended without Changing the Centers[,] for it plainly appears that the Center of the Plummet and the Center of the Telescope is not the Same and p[er]haps neither of them the true Center of the Graduations so that the Instrum[en]t cannot be de-pended on to come w[i]th certainty within less than 5 Min[u]t[e]s[.] Could I be sure of an Instrum[en]t

[43] Logan is not clear, but he may have referred to James Pound (1669–1724), the English mathematician who had a private ob-servatory at Wanstead where he worked with his nephew James Bradley who later became the Astronomer Royal. In 1719 Pound helped Bradley construct a clockdriven telescope. (E. G. R. Taylor, op. cit., p. 139.)

that would in all cases come within one Minute I should not think much of the Money but to lay down so much & still be at an Uncertainty is not worth while.

I have been thinking of getting a large Sector of very well Seasoned Box with some plates of brass to keep it from warping of about 30 Inches long with a graduated 3d legg to make it a Triangular Quadrant well Secured with brass Screws as well socket as pin, the Center of the Graduations to be the Center of the Sector & therefore to make the principal part of a Sectant only and the graduations to be by Diagonals by ten parallels to a minute w[hi]ch the size of the Instrum[en]t will well enough admitt of also to have another Center as the common Quadrant Sectors have near the Middle of one of the legs to take in a Whole Quadrant & therefore to serve but less exactly to 90 deg[rees] on a Single Line. This to be fitted with Telescope Sights & that the whole be taken to pieces & stowed into a Portmanteau to be used anywhere in the woods where full Instrum[en]ts could not be used because of their bulk whereas any Traveller might carry one of these unobserved. It would want also some Apparatus to fix it. Pray consult thy ffriend Gwyn[44] What the charge of such an Instrum[en]t might be. The Merid[ian] Alt[itude] of the Sun is always in these parts of the World above 30 degr[ees] except in Winter and that is a Season people are not fond of travelling in the Woods to make observations[.]

"ON THE INVENTION OF THE QUADRANT, COMMONLY CALLED HADLEY'S," BRADFORD'S *American Magazine*, JULY, 1758.[45]

The great improvement which the art of navigation has received from the invention of this instrument, must ever place those concerned in it, among the highest class of names that will be remembered by posterity. Tho' Mr. Hadley, (whose fame in the learned world can suffer no dimunition by what we are now to publish) has great merit in the improvement of this instrument which bears his name, yet there is sufficient reason to conclude that he was not the first inventor.

In the philosophical transactions, No. 435 there is an "Account of Mr. Thomas Godfrey's Improvement of Davis's Quadrant transferred to the Mariner's Bow" by the late Mr. Logan of this place, whose reputation in mathematics was inferior to few in his day. In that account it appears that Mr. Godfrey, of this city, had begun to think of this matter as early as the year 1730. He was a glazier by trade, and a man of

---

[44] Not identified. The context indicates that Gwyn was probably a maker of navigational instruments.

[45] This posthumous publication of two letters by James Logan on the invention of the sea-octant reveals the continuing interest in the controversy years after Logan's death. Logan's letter to Halley, printed in the *American Magazine*, should be compared with Logan's article in No. 435 of the *Philosophical Transactions* of the Royal Society of London.

no education, but perhaps the most singular phenomenon that ever appeared in the learned world, for a kind of natural or intuitive knowledge of the abstrusest parts of Mathematics and Astronomy.

In order, therefore, to shew how far the honor of this invention is due to Mr. Godfrey and his patron Mr. Logan, we propose to publish Mr. Logan's original account referred to in the above quoted number of the philosophical transactions; together with two letters to the royal society written previous to that account, one by Mr. Logan & the other by Mr. Godfrey himself. These three letters will give a complete view of the whole affair in its rise and progress. They were put into our hands by a sensible and candid citizen of Philadelphia, with the following pertinent introduction, and, therefore, their authenticity, if it were doubted, may be easily vouched.

To the Proprietors &c.

Gentlemen

All civilized states have thought it their honor to have men of great ingenuity born or bred among them. Many cities of ancient Greece had long and sharp contentions for the honor of Homer's birth-place. And, in later times, volumes have been written in Europe in disputing which city had the true claim to the invention of the art of printing. Nor is to be wondered that mankind should be so generally eager in this respect, since nothing redounds more to the honor of any state than to have it said that some science of general utility to mankind was invented or improved by them.

Nevertheless, it often happens that the true authors of many an useful invention, either by accident or fraud lose the credit thereof; and, from age to age, it passes in the name of another. Thus it happened heretofore to Columbus and many others; and thus also it has happened to a native of Philadelphia.

Mr. Thomas Godfrey, it is well known to many of us here, was the real inventor of that very useful instrument called Hadley's quadrant, or octant. To him the merit is due, and to his posterity the profit ought to belong. This, will fully appear from the three following genuine letters, which, I persuade myself you'll think worthy of being recorded in your magazine, in order to restore, as far as possible, the credit of that invention to our city, and to the posterity of Mr. Godfrey. How he came to be deprived of it may be made a question by some. I answer that Mr. Godfrey sent the instrument to be tried at sea by an acquaintance of his, an ingenious navigator in a voyage to Jamaica, who shewed it to a captain of a ship there just going for England, by which means it came to the knowledge of Mr. Hadley, tho' perhaps without his being told the name of the real inventor. This fact is sufficiently known to many seamen and others yet alive in this city; and established beyond doubt by the following letters written about that time.

It is therefore submitted to the world whether, after perusing the letters, they ought not in justice to call that instrument for the future Godfrey's, and not Hadley's, quadrant?

To Dr. Edmund Halley.

Esteemed Friend

The discovery of the longitude having of late years employed the thoughts of many, & the world now expecting from thy great sagacity and industry some advances towards it far exceeding all former attempts, from the motion of the moon; to the ascertaining of which thy labours have so long, and happily been directed; the following notice, I hope will neither be thought unseasonable nor prove unacceptable. That the success of that method depends on finding the moon's true place for one meridian by calculation and for another by observation, I think is generally allowed: The first of which being depended on from thy great genius, what remains is some certain method for observation, practicable on that unstable element the sea.

In order to this thy predecessor at Greenwich [John Flamsteed], if I mistake not, for some years published his calculation for the moon's future appulse to the fixt stars, which would save all observation, but that of a glass: but these not often happening, and the moon often having a considerable parallax when they did that project dropt.

For finding her place by taking her greater distances from stars, the forestaff or crossstaff cannot be exact enough: And quadrants sextants &c. with two telescopes are impracticable at sea.

Dr. Biester's late proposal for taking the difference of rad: ascension between the moon and a star, if that should prove practicable with sufficient exactness would undoubtedly answer the intention of all that is to be expected from the moon, if her place were taken on or near the meridian. But to keep the arch of this instrument in the plane of the equator, and at the same time two objects of unequal altitudes and considerable distance from each other, by the edges of two sights, with the necessary accuracy, will not perhaps be so easy in practice as he would have it believed.

I shall therefore here presume from thy favour shewn me in England, in 1724 to communicate an invention, that whether it answer the end or not, will be allow'd, I believe to deserve thy regard. I have it thus.

A young man born in this country Thomas Godfrey by name, by trade a glasier, who had no other education, than to learn to read and write, with a little common arithmetick, having in his apprenticeship, with a very poor man of that trade, accidentally met with a mathematical book, took such a fancy to the study, that by the natural strength of his genius, without any instructor, he soon made himself master of that, and of every other of the kind, he could borrow or procure in English: And finding, there was more to be had in Latin books, under all imaginable discouragements, applied himself, to the study of that language, till he could pretty well understand an author on these subjects, after which, the first time I ever saw or heard of him, to my knowledge, he came to borrow Sir Isaac Newton's principia of me. Inquiring of him hereupon who he was, I was indeed astonished at his request, but after a little discourse, he soon became welcome to that or any other book I had. This young man about 18 month's since, told me had for some time been thinking of an instrument for taking the distances of stars by reflecting speculums, which he believed might be of service at sea, and not long after, he showed me a common sea quadrant, to which he had fitted 2 pieces of looking glass, in such a manner as brought two stars at almost any distance, to coincide, the one by a direct, the other by a reflected ray, so that the eye could take them both together, as joined in one, while a moving label or index on the gratuated [graduated] arch, marked exactly half their distance. For I need not say, that the variations of the angels [angles] of reflection, from two speculums are double to the angle of the inclination of their plains [planes], and therefore gives but half the angle or arch of the distance, which is the only inconveniency that appears to me to attend this. But as it may be made so simple, easy and light, as not to be much more unwieldy or unmanageable, tho' of a considerable length, than a single telescope of the same, that inconveniency will be abundantly compensated.

The description of it, as he proposes it, and has got one made, is nearly thus, which he is willing I should communicate to thee if possibly it may be of service.

To a streight ruler or piece of wood AB of about 3 inches in breadth, and from 40 to 45 in length, (or of any other that may be thought convenient) with a suitable thickness, an arch or limb AC of about 30 degrees to the radius KL is to be fixed. To the upper end of the piece AB, a piece DD is to be morticed, and, in it the center K taken, so that OP may be about 6 inches, and the angle KOP about 40 degrees. On this center K the ruler in index KL is to move having a fiducial edge below answerable to the central point to cut the graduations on the limb. On the upper end of this index a speculum of silvered glass or rather metal exactly plain EF, of about 3 inches in length and two in height is erected perpendicular to the plane of the index, and also nearly at right angles with its sides, the plane of the reflecting surface standing exactly over the central point. At the end B of the piece AB another speculum of glass is to be in the same manner erected which may be somewhat less than the other, with a square or oblong spot in it unsilvered, that a star by a direct ray may be seen thro' it, and the back of this speculum should be guarded with a thin brass plate, with an aperture in it equal to the unsilvered part of the glass, the edge of the aperture toward *H* to be exactly streight dividing between the silvered and unsilvered part of the speculum, and standing in the line of the axis of the telescope. This speculum is to be set at an angle, of about 20 deg. with the square of the piece *AB*, or at 110 degrees with the sides of it. Upon the piece *AB* the telescope *PQ* is fixed of a good aperture and field, with the axis placed as above. The limb is to be graduated by diagonals or parallel circles, to half degrees and half min. beginning from *C* which are to be numbred as whole ones. And if it be practicable to face wood with brass without warping, the whole face should be so covered; if not, then along the outward edge of the limb, a narrow strip of brass plate may be let into the face of it, finely and equally indented on the edge, to take a screw fitted to that toothing to be fixed on the moving index as *L* as your instruments are made, that count by revolutions; and then before this is used, it would be proper to take the distance of the two objects first nearly by a forestaff, and from thence accordingly to set the index. This screw at land would be highly usefull, but at sea it cannot be wrought, while the instrument is directed by the same person, tho' as the motion's [*sic*] of the moon and variation of the angle is but slow, it may be brought to exactness by several trials in the intervals of direction. The instrument as above described, will not take an angle of much above 50 degrees, which for the purpose intended may be fully sufficient. But if the speculum *EF* be made to take off and put on, and the end of the index at *K* be so notched as to turn that speculum from its first perpendicularity, to make an angle of about 25 degrees, it will then take any distance to 100 degrees.

By this description it may be thought that the utmost accuracy will be required in making the instrument. Yet of all that ever have been invented of so curious a kind it will probably be found to demand the least; for provided the speculum's [*sic*] are good, on which the whole depends, if the first *EF* be set truly over the center, the limb well graduated and the other speculum be also set perpendicular, there can (I think) be no other error but what the instrument it self will easily rectify: for if it be directed to one star, and that be taken at the same time both by a direct ray thro' the glass *GH*, and by a reflection from *EF*, both exactly coinciding at *O*, 'tis evident that then the speculum's [*sic*] are exactly parallel. And if this fall not precisely when the index cuts 0 degrees, if the variation be noted, this constantly added or subtracted according as it falls, will fully rectify all other errors. So in fixing the speculum *EF* to another angle, as has been proposed, the like method may or must be taken viz. to observe 2 stars at the distance of about 45 or 50 degrees by the speculum in it's [*sic*] first situation, and then the same stars by it again in it's [*sic*] 2d. and the difference of the intersection of the index on the limb being noted and constantly added to the

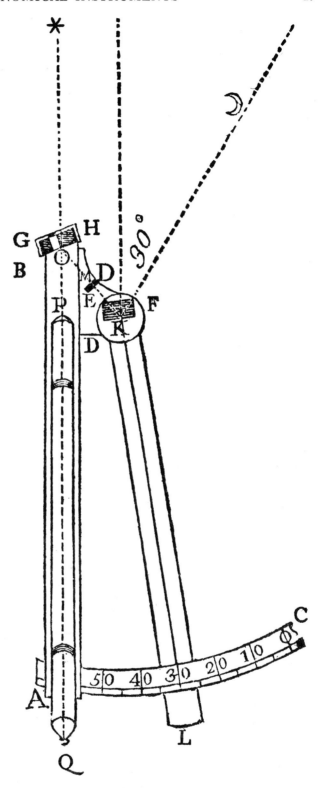

arches taken in the 2d situation, will give the true distance. This method of observing one and the same star, in the first manner, or 2 stars in the second as had been mentioned, will also rectify errors even in the speculums: for the line of the ray *KO* is in all cases constantly the same, and upon the whole I may safely say the instrument will be found much more certain in practice than at first it may appear in theory, even to some good judges. But I am now sensible I have trespassed in being so particular, when writing to Dr. Halley, for I well know that to a gentleman noted for his excellent talent of reading, apprehending, and greatly improving, less would have been sufficient; but as this possibly may be communicated by thee, I shall crave leave further to add, that the use of the instrument is very easy. For if the index be set so near the distance of the moon and stars, and the limb so held as to cut the body of the moon, upon directing the telescope to the star, her image will of course be reflected on some part of the speculum *GH*. There is no absolute necessity, the star and moon should coincide exactly at the line limiting the silvered and unsilvered part of the latter speculum; for the transparent part of that glass will often reflect the moon's image sufficiently for the telescope to take it, and if her limb in that and the star exactly coincide near it, it may be sufficient, tho the nearer to that line the better. Now their distance being found, the tables that give the moon's place may be depended on for her diameter and her latitude, which last being known there are 3 sides of a triangle given to find the angle at the pole of the ecliptic, which compared with the star's longitude, determine her place for that instant; for in respect to her lattitude, when she is swiftest in motion when nearest her nodes, and when the inclination of the orb is greatest (if these could all happen together) yet the variation of her latitude, in the space of one hour equal to 15 deg: of longitude on the earth, if a star be taken somewhat near the ecliptic and not very near the moon, will not alter the angle at the pole but a very few seconds. The nearness of the speculum *GH* is no disadvantage, because the rays come reflected in the same manner as they come direct. It may be needless to add that, when practicable, the moon should be taken when near the meridian;—or that the instrument will equally take the distance of the sun from the moon when visible, as she often is in the day time; for which purpose, there must be a place made at M for a darkning glass, to be fixed there when necessary, and the telescope directed to the moon. Nor need I add that the same instrument will very well serve for taking the distance of any two stars, a comet &c. always taking the brightest by reflection, all which is obvious. But I must further observe with pleasure, that if we do not quite mistake in all that has been said here, there is now a method found by it to obtain, what is equivalent to a bodily appulse of the moon to a fixed star, or to the sun at any moment when visible, which indeed might be wished, but could scarce be hoped for by any means to be used at sea; and therefore, if the longitude could ever be expected to be determined by the motions of the moon, (to which end I. Flamstead's and thy more assiduous

labours in observing her, have I suppose been principally levelled) and this instrument be duly made to answer what is proposed, as it may be framed light and easily manageable, thou wilt then with thy accurate tables, have obtained the great desideratum and all that can in this way be had from our satellites. And if the method of discovering the longitude by the moon is to meet with a reward, and this instrument, which for all that I have ever read or heard of, is an invention altogether new, be made use of, in that case I would recommend the inventor to thy justice and notice. He now gets his own and family's bread (for he is married) by the labour of his own hands only, by that mean trade. He had begun to make tables of the moon, on the very same principles with thine, till I lately put a copy of those that have lain so many years printed but not published with W. Inny's, into his hands, and then highly approving of them, he desisted. We both wish very much to see thy tables compleated, and ushered into the world by thy own hand. On thy receipt of this I shall hope for a line, with thy thoughts on it; which however they prove will afford a pleasure to thy real friend.

J. Logan.

Pennsylvania, May 25, 1732.

LOGAN TO WILLIAM JONES, PHILADELPHIA, NOVEMBER 8, 1732, RIGAUD 1: pp. 282–283.

I must with gratitude acknowledge thy obliging present of three little books, by thy quondam pupil J. Georges,[46] though a line from thy hand would have rendered them still more acceptable. I could not, on sight of them, but very much admire at those great improvements by J. Machin[47] on Sir Isaac's theory of the moon, now published so soon after that great man's death, and not mentioned before, which, had they been early enough communicated to so industrious a hand, might perhaps have been formed into that method and order, fortified with proper demonstrations, which they at present seem to want. I am willing to impute it to my own incapacity, that I can by no means comprehend his law of motion, where a body is deflected by two forces tending constantly to two fixed points, viz. that it will describe, by lines drawn from the two fixed points, equal solids, in equal times, about the line joining the said fixed points, which are his words, but such as I can, by no construction of them, form a notion of a solid that will duly increase in quantity in any wise proportionally to the times of description: nor can J. Georges, though he

[46] Not identified.

[47] John Machin, an English astronomer, acted as secretary of the Royal Society from 1718 to 1747, and he was professor of astronomy at Gresham College from 1713 to 1751, the year of his death. His attempt to improve upon Newton's lunar theory was published as "Laws of the Moon's Motion according to Gravity," an appendix to Motte's translation of the *Principia Mathematica* (London, 1729).

appeared at first acquainted with the thing, give me any light into it. A young man here, also, of an excellent natural genius for these studies, who, under the greatest disadvantages of education and circumstances, has made himself a very good master of the Newtonian philosophy, and that great author's writings, finds himself as much at a loss in it as I do: and yet perhaps one line might render it intelligible to us. But I cannot pretend a right to desire such a favour, however obliging it might prove.

That young man, having not long since invented an easy and curious instrument for taking at sea the moon's real place in the heavens, by her distance from any known fixed star near the ecliptic, though 50 or 60 degrees distant from the moon, and this very near, if not altogether, as exactly as it could be done by a real transit, which is performed by means of one small speculum, with a much smaller visual passage through it, fixed a few inches before the further end of a short telescope of 30 or 36 inches, and another such small speculum but wholly reflective, fixed at the end of a label or index, of about the same length with the telescope, moving on a centre exactly under the reflecting surface of the speculum erected on it, and by its fiducial edge below marking the graduations, accounting one degree two, on a short limb at the lower end of the piece, on which the telescope is fixed; that young man, I say, having invented this, and shewn me one he had made, believing it might prove of great use for the purpose above mentioned, in May or June last I sent Dr. Halley an account of it, with a figure and description of the instrument to be applied by him as he should think proper. How it may appear to his better judgment, I cannot guess; but if thou please to make some little inquiry into it, in a proper manner, and favour me with a line upon this letter, I shall be highly obliged to thee, which to save trouble may be directed to me, in a cover to Lawr. Williams, merchant, and left at the Pensilvania Coffee-house in Birchin Lane, from whence he will carefully forward it to thy sincere and most obliged friend,

J. Logan.

10th Nov. P.S. Since the above, Tho. Godfrey, the young man I mentioned, has, I think hit on the meaning of that rule of J. Machin; but, according to his account of it, 'twill be of no use when the body moves in the same plane with the centres; for then there can be no solid. I suppose he will take the freedom to write to thee on another subject. He wishes I had directed my letter, with his invention of the reflecting instrument, to thee, instead of Dr. Halley.

LOGAN TO JONES, PENSILVANIA, NOVEMBER 12, 1734, RIGAUD 1: pp. 284–287.

About this time two years, by Capt. Wright, I returned thee my thanks for thy kind present of two or three small pieces delivered me by J. Georges. Since which I have been informed by my friend P. Collinson and said Wright, of thy generosity and justice in asserting before the Royal Society, the right of an inventor (at least), if not absolutely the first, of the reflecting instrument, to Tho. Godfrey, as well as in vindication of my reputation from the slur, that Dr. Halley's unhandsome conduct towards me had like to have thrown on it, in which he was highly ungrateful; since nothing but my respect for him could induce me to communicate to him, preferably to all others, what he might easily judge, from my letter, I thought would be wholly new to him. And to suspect a trick or sham in it, he must have considered me as one of the most senseless or maddest creatures upon earth, if I should voluntarily, in so wild a manner, expose myself even to a hazard of the vile imputation of an impostor in a matter, wherein I proposed to myself neither credit nor profit, nor any advantage whatsoever. Thy kind endeavours, therefore, to obtain justice both to T. G. and me deserve my hearty acknowledgments, which I here take the freedom to make thee; and I should do the same to the ingenious J. Machin, to whom I find I was also particularly obliged, but not having the honour of any acquaintance with him, I request thee, when you meet, to do it for me in my name.

Being told you had wished for a fuller account of T. Godfrey's improvement of the mariner's bow, than he had himself given of it, which he sent without my knowledge, I took the trouble of drawing up one, this last summer, and sent it to P. Collinson, which, if he has received and communicated it, I hope will give satisfaction, as I also do, that T. G.'s merit in the other instrument will not be forgotten; for as that instrument of his was not only made, but used at sea six months before J. Hadley's was seen or known, and my description of it will, I suppose, be allowed to be much plainer, and the use of it applied to nobler purposes in my letter, T. G. justly deserves the preference. To have sent a demonstration of the principle, on which the instrument is formed, to Dr. Halley, would have been very needless; but I drew one myself, which if I can find it, and have time, I shall send thee, that if I mistake not, will appear somewhat clearer than J. H.'s. I thought also to have mentioned to thee some other mathematical subjects, which at present I shall defer, and conclude this with sincere respect from

thy much obliged friend,
J. Logan.

P.S. The ship staying longer than expected, I have not only sent the demonstration mentioned above, but another that I have now struck out, which I take to be preferable to all others, for the reasons given at the close of it.

J. L.

[Rigand here omitted Logan's demonstration "that the angle made with one another by the reflecting planes is half of that which will be contained by the directions of the incident and reflected ray."]

The preceding demonstration, I suppose, will be allowed strictly geometrical; but to those, who require not such, this other may appear more plain and satisfactory, as it not only sufficiently proves the proposition, but at the same time also shows the reason, and that it must be so.

Let the glasses AB and CD (Plate 3, fig. 1)[48] be supposed to be exactly parallel, and the ray IE, falling on AB, make with it the angle IEA, which being reflected from E on the glass CD in the line EF, and again from F in Ff, makes the four angles IEA, BEF, CFE, and DFf, all equal, and Ff and IE are parallel, or the ray IE continued in the same direction, and let each of these angles be called $z = 1$.

Let the glass AB be moved from its parallel position into that of aEb, making with its first (now the line AB) the angle $aEA = 1/4z =$ the angle bGD, which reduces $z = 1$ to $3/4z =$ the angle IEa, and the progress of the ray IE will now (by Dioptr.) be IE, EH, HK. And because the angle $HEb =$ the angle $IEa = 3/4$, because the angle bEB (part of it) $= 1/4$; therefore the angle $HEB = 2/4$ only. Let Hh be drawn parallel to IE, then because AB and CD are parallel, the angle DHh is [equal to] the angle $IEA = z = 1$, and for the same reason the angle $BEH =$ the angle $CHE =$ the angle $DHK = 2/4$. But the angle DHh ($=1$) $-$ the angle DHK ($= 2/4$) is $= 2/4 =$ the angle $KIE =$ (because the angle $bGD =$ the angle $AEa = 1/4$) $=$ twice the angle bGD. Q.E.D.

A bare view of the figure, and of the progress of the ray IE in the difference positions of the glass AB, ab, will abundantly shew the reason of this; and therefore, though the first demonstration is more agreeable to the strict method of the ancients, yet this for instruction is preferable. T. Godfrey also shewed me his, which was a good one; but as it was neither altogether so geometrical as the first of these, nor so clear as the last, I have now taken the freedom to send these two of my own.

AN ACCOUNT OF MR. THOMAS GODFREY'S IMPROVE-MENT OF DAVIS'S QUADRANT, TRANSFERRED TO THE MARINER'S-BOW, COMMUNICATED TO THE ROYAL SOCIETY, BY MR. J. LOGAN, *Phil. Trans.* **37**, Num. 435: pp. 441–450.

Being inform'd that this Improvement, proposed by Thomas Godfrey of this Place, for observing the

Sun's Altitude at Sea, with more Ease and Expedition than is practicable by the common Instruments in use for that purpose, was last Winter laid before the Royal Society, in his own Description of it; and that some Gentlemen wish'd to see the Benefit intended by it more fully and clearly explained: I, who have here the Opportunity of knowing the Author's Thoughts on such Subjects, being perswaded in my Judgment that if the Instrument, as he proposes it, be brought into Practice, it will in many Cases be of great service to Navigation, have therefore thought it proper to draw up a more full Account of it, than the Author himself has given, with the Advantages attending it; which if approved of by better Judgments, to whom what I offer is entirely submitted, 'tis hoped the Use of it will be recommended and further encouraged, as also the Author. The Rise of the Improvement with its Conveniencies, as also a Description of it, are as follows.

Tho. Godfrey having under the greatest Disadvantages (as I observed in my first Letter to Dr. Halley, giving an Account of his Invention of the Reflecting Instrument) made himself Master of the Principles of Astronomy and Optics, as well as other Parts of Mathematical Science, applied his Thoughts to consider the Instruments used in that most momentous Part of Business, Navigation. He saw that on the Knowledge of the Latitude and Longitude of the Place a Ship is in, the Lives of thousands of useful Subjects, as well as valuable Cargoes, continually depend; that for finding the first of these, certain and easy Methods are furnish'd by Nature, if Observations be duly made: But Davis's Quadrant, the Instrument generally used by British Navigators, (tho' seldom by Foreigners) he perceived was attended with this Inconveniency, that the Observer must bring the Shade or Spot of Light from the Sun, and the Rays from the Horizon, to coincide exactly on the fiducial Edge of the horizontal Vane: That tho' this can be done in moderate Weather and Seas with a clear Sky, and when the Sun is not too high, without any great Difficulty; yet in other Cases it requires more Accuracy than can in some Junctures possibly be applied, and more Time than can be allowed for it. In European Latitudes, or to those nearer the Northern Tropick, when the Sun is in the Southern Signs, and near the Meridian, he rises and falls but slowly: Yet in Voyages to the East and West-Indies, of which very many, especially to the latter, are made, he is at Noon, often and for many Days together, in or near the Zenith, when he has Declination, faster than even at the Horizon; for it is well known to Persons acquainted with the Sphere, that when his diurnal Course takes the Zenith, he there rises and falls a whole Degree or 60 Minutes, in the Space of four Minutes of Time; so that the Observer has but one Minute, to come within 15 Minutes of the Truth in his Latitude: While in a middle Altitude, as 45 Deg. he is at Noon about 5 Minutes and a half in Time, in rising or falling one

---

[48] Logan's first demonstration is not given in Rigaud. However, Rigaud summarizes it briefly as follows: "Logan's first method is the same as is now to be found in all books of optics, where it is demonstrated that the angle made with another by the reflecting planes is half of that which will be contained by the directions of the incident and reflected ray." (Rigaud, **1**: p. 286.) Logan's reference to plate 3, fig. 1 is unclear.

single Minute of Space, the Odds between which is more than 80 to 1. And yet perhaps, no Parts of the World require more Exactness in taking the Latitude than is necessary in Voyages to the West-Indies: For it is owing to the Difficulty of it, that Vessels have so frequently miss'd the Island of Barbados, and when got to the Leeward of it have been obliged to run down a thousand Miles further to Jamaica, from whence they can scarce work up again in the Space of many Weeks, against the constant Tradewinds, and therefore generally decline to try for, or attempt it.

But farther, as the Latitude cannot be found by any other Method, that our Mariners are generally acquainted with, than by the Sun or a Star on the Meridian: In a cloudy Sky, when the Sun can but now and then be seen, and only between the Openings of the Clouds for very short Intervals, which those who use the Sea know frequently happens: As also in high tempestuous Seas, when tho' the Sun should appear, the Observer can scarce by any Means hold his Feet; it would certainly be of vast Advantage to have an Instrument by which an Observation could also be, as it were, snatched or taken in much less Time, than is generally required in the Use of the common Quadrant.

Tho. Godfrey therefore considering this, applied himself to find out some Contrivance by which the Necessity of bringing the Rays from the Sun, and those from the Horizon to coincide (which is the most difficult part of the Work) on one particular Point or Line from the Centre, might be removed. In order to which he consider'd, that by the 21. 3d Elem. of Eucl.[49] all Angles at the Periphery of a Circle, subtended by the same Segment within it are equal, on whatever part of the Circumference the angular Point falls; and therefore, if instead of a Quadrant, a Semicircle were graduated into 90 Degrees only, accounting every two Degrees but one; this would effectually answer: For then, if an Arch of the same Circle were placed at the End of the Diameter of the Instrument, every Part of the opposite Arch would equally serve for taking the Coincidence of the Rays above-mentioned. But such an Instrument would manifestly be attended with great Inconveniencies; for it would in great Altitudes be much more unmanageable, and the Vanes could not be framed to stand, as they always ought, perpendicular to the Rays. He therefore further resolved to try whether a Curve could not be found to be placed at the Centre of a Quadrant, which would, at least for a Length sufficient to catch the Coincidence of the Rays, with Ease fully answer the Intention.

A Curve that in all the Parts of it would in geometrical strictness effect this, cannot be in Nature, any more than that one and the same Point can be found for a Centre to different Circles, which are not concentric. It is certain that every Arch on the Limb may have a Circle that will pass through the Center,

49 The reference is to Euclid, *Elements*, Book 3, Proposition 21.

and be a Locus or geometrical Place for the Angle made by that Arch to fall on: but then every Arch has a different one from all others; as in the Figure. Let *ABC* be the Quadrant, and *AB*, *EF*, *GH* be taken as Arches of it: Circles drawn through each two of these respectively, and through the Center *C* as a third Point, will manifestly be such Loci or Places: For every Pair of these Points stand in a Segment of their own Circle, as well as on a Segment of the Quadrant; and therefore by the cited 21. 3d Elem. the Angles standing on these first Segments will every where be equal at the Periphery of their respective Circles, and their Radius will always be equal to half the Secant of

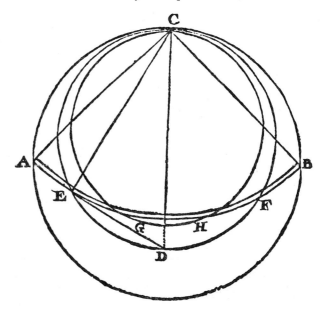

half the Arch on the Quadrant. For in the Circle *CEDF* (for instance) the Angle *CED* is right, because 'tis in a Semicircle, *CE* is the Radius of the Quadrant, *ED* the Tangent of the Angle *DCE* = 1/2 the Arch *EF*, and *CD* is the Secant of the same = the Diameter of the Circle *CEDF*, and therefore its Radius is half that Secant.

Now from the Figure 'tis plain, that in very small Arches the Radius of their circular Place will be half the Radius of the Quadrant, that is, putting this Radius = 10, the other will be 5. And the Radius for the Arch of 90, the highest to be used on the Quadrant will be the Square Root of half the Square of the Radius = Sine of 45 Degrees = 7.071, and the Arches at the Center drawn by these two Radii are the Extreams, the Medium of which is 6.0355. And if a circular Arch be drawn with this Radius 1/20 the Part of the Length of it, that is, in an Instrument of 20 Inches Radius, the Length of one Inch on each Side of the Center affording two Inches in the whole, to catch the Coincidence of the Rays on, which must be own'd is abundantly sufficient, the Error at the greatest Variation of the Arches, and at the Extremity of these 2 Inches, will not much exceed one Minute.

But in fixing the Curvature or Radius of this Central Arch, something farther than a Medium between the Extreams in the Radius is to be considered: For in small Arches the Variation is very small, but in greater it equally encreases, as in the Figure where it appears, the Difference between the Angles *ABC* and *ADC* is much greater than the Difference between *EBC* and *EDC*, though both are subtended by the same Line *BD*; for their Differences are the Angles *BAD* and *BED*. Therefore this Inequality was likewise to be considered; and compounding both together, Tho. Godfrey pitched on the Ratio of 7 to 11, for the Radius of the Curve to the Radius of the Instrument, which is 6.3636 to 10. But on further Advisement he now concludes on 6-4/10; and a Curve of this Radius of an Inch on each Side of the Center to an Instrument of 20 Inches Radius or of 11/20th of the Radius, whatever it be, will in no Case whatever, as he has himself carefully computed it, produce an Error of above 57 Seconds; and 'tis very well known that Navigators (as they very safely may) in their Voyages intirely slight a Difference of one Minute in Latitude.

This Radius is the true one for the circular Place to an Arch of 77°15', and the Variation from it is nearly as great at 90 Degrees as at any Arch below it, the greatest below being at about 44 Degrees, which is owing to the Differences expressed by the last Figure above, and not to those of the Curvatures or circular Places. Yet this Variation of 57 Seconds arises only when the Spot or Coincidence falls at the Extremity of the horizontal Sight or Vane, or a Whole Inch (in an Instrument of 20 Inches Radius) from the Center, and then only in the Altitudes or Arches of about 44 or 90 Degrees. And in these, at the Distance of half an Inch from the Center, the Variation is but 1/4 so much, viz. about 14"; and at 1/4 of an Inch, not 4"; at the Center tis precisely true. Therefore as an Observation may be taken with it in one fourth of the Time, that Davis's Quadrant, on which three Things must be brought to meet, in a general way requires: I say, considering this, and the vast importance of such Dispatch, in the Case of great Altitudes, or of tempestuous Seas, or beclouded Skies, tis presumed the Instrument thus made will be judged preferable to all others of the kind yet known. Some Masters of Vessels, who sail from hence to the West-Indies, have got of them made as well as they can be done here; and have found so great an Advantage in the Facility and in the ready Use of them, in those Southerly Latitudes, that they reject all others. And it can scarce be doubted, but when the Instrument becomes generally known, it may upon the Royal Society's Approbation, if the Thing appear worthy of it, more universally obtain in Practice.

'Tis now four Years since Tho. Godfrey hit on this Improvement; for his Account of it, laid before the Society last Winter, in which he mentions two Years, was wrote in 1732. And in the same Year, 1730, after he was satisfied in this, he applied himself to think of the other, viz. the reflecting Instrument by Speculums, for a help in the Case of Longitude, though 'tis also useful in taking Altitudes, and one of these, as has been abundantly proved by the Maker, and those who had it with them, was taken to Sea, and there used in observing the Latitude, the Winter of that Year, and brought back again hither before the End of February, 1730/1, and was in my keeping for some Months immediately after. It was unhappy indeed, that having it in my Power, seeing he had no Acquaintance nor Knowledge of Persons there, that I transmitted not an Account of it sooner: But I had other Affairs of more Importance to me: And it was owing to an Accident which gave me some Uneasiness, viz. his attempting to publish some Account of it in Print here, that I did it at that Time, viz. in May 1732, when I transmitted it to Dr. Halley; to whom I made not the least Doubt but the Invention would appear entirely New. This, on my part, was all the Merit I had to claim, nor did I then, or now assume any other, in either of these Instruments. I only wish that the ingenious Inventor himself might by some means be taken Notice of, in a Manner that might be of real Advantage to him.

There needs not, I suppose, much more of a Description of the Instrument than has been given: I shall only say, the Bow had best be an Arch of about 100 Degrees well graduated, and numbered both ways; the Radius of 20 or 24 Inches; the Curve at the Center to be 1/20th of the Radius on each Side, that is, 1/10th of it in the whole; the Radius of that Curve 64/200 Parts of the Radius of the Instrument; that the Glass for the Solar Vane should not be less, but rather larger, than a silver Shilling, with its Vertex most exactly set. And that the utmost Care be taken to place the Middle of the Curve at the Centre exactly perpendicular to the Line or Radius of 45 Degrees. As the Observer must also take Care that the two Vanes on the Limb be kept nearly equidistant from that Degree; to which I shall only add, that it may be best to give the horizontal Vane only one Aperture, and not two. The rest I suppose may be left to the Workmen. Thus doubting I have already been too prolix on the Subject, to which nothing but a sincere Inclination to promote any thing that might contribute to a publick Benefit, and to do some justice to Merit, could induce me, I shall only request that what I have here offered may be construed by that Intention.

Philad. 28th of June, 1734   J. Logan

Note, That the Radius of the Quadrant being divided into 20 equal Parts, the Center *X* (*in fig. 1*) of the Curvature of the Horizon-Vane (*AB*) must be 12-8/10 of those Parts from the Center (*C*) of the

Quadrant. The Breadth (*AB* or *gh*) of that Vane should be 1/10 of the whole Radius, that is 1/20 on each Side of the Center (*C*).

AN EXTRACT OF A LETTER FROM JAMES LOGAN, ESQ; TO SIR HANS SLOANE, BART. R. S. PR. CONCERNING THE CROOKED AND ANGULAR APPEARANCE OF THE STREAKS, OR DARTS OF LIGHT'NING IN THUNDER-STORMS, *Phil. Trans.*, Num. 441: p. 240.

Philadelphia, September 20, 1735

I shall crave thy indulgence further to add, that looking into the Second Volume of that Ingenious Gentleman Stephen Hales's Staticks,[50] of which I have lately happened to have a Cursory View, I observed him to mention that Phaenomenon of the Streaks or Darts of Light'ning in Thunder-Storms appearing crooked and angular (I do not remember his Words) as a thing not yet accounted for; and therefore he guessed at a Solution of it.

The Clouds are generally distinct Collections of Vapours like Fleeces, and therefore the Rays of Light thro' them must pass thro' very different Densities, and accordingly suffer very great Refractions; From thence therefore, undoubtedly, that Appearance must arise. For it is most highly absurd to imagine, that Fire darted with such a Rapidity, can from any assignable Cause deviate in Fact from a Right Line, in the Manner it appears to us. And this, if duely considered, may probably be found a plenary Solution.

SOME THOUGHTS CONCERNING THE SUN AND MOON, WHEN NEAR THE HORIZON, APPEARING LARGER THAN WHEN NEAR THE ZENITH; BEING PART OF A LETTER FROM JAMES LOGAN, ESQ; TO SIR HANS SLOANE, BART. PRESIDENT OF THE ROYAL SOCIETY, AC, *Phil. Trans.*, Num. 444: p. 404.

Philadelphia, Sept. 20, 1735

It may, perhaps, be needless now to add any thing in Confirmation of Dr. Wallis's Solution (See these Transactions, No. 187.)[51] of the Sun and Moon's appearing so much larger at rising or setting, than when in a greater Altitude; tho' some have very absurdly still gone on to account for it from Vapours, which I remember was given to me in my Youth for the true Cause of it. 'Tis true, indeed, that 'tis these Vapours, or the Atmosphere, alone, that make those Bodies, when very near to the Horizon, appear in a spheroidal Form, by refracting, and thereby raising (to Sight) the lower Limb more than the upper yet these can be no Cause of the other. The Sun and Moon, each subtending about half a Degree, appear in the Meridian of the Breadth of eight or ten Inches,

to some Eyes more, and to others less; and in the Horizon to be two or three Foot, more or less, according to the Extent of Ground they are seen over: But if one can have an Opportunity, as I have here frequently had, of seeing the Sun rise or set over a small Eminence at the Distance of a Mile or two with tall trees on it standing pretty close as is usual in Woods without Underwood, his Body will then appear to be ten or twelve Foot in Breadth, according to the Distance and Circumstances of the Trees he is seen through; and where there has been some thin Underwood, or a few Saplings, I have observed that the Sun setting red, has appeared through them like a large extensive Flame, as if some House were on Fire beyond them. Now the Reason of this is obvious, viz. that being well acquainted with Trees, the Ideas of the Space they take up are in a manner fix'd, and as one of those Trees subtends an Angle at the Eye, perhaps not exceeding two or three Seconds, and would scarce be distinguishable, were it not for the strong Light behind them, the Sun's Diameter of above thirty Minutes takes in several of them, and therefore will naturally be judged vastly larger. Hence 'tis evident, that those Bodies appear greater or less, according to the Objects interposed or taken in by the Eye on viewing them. And to this only is that Phaenomenon to be imputed.

I am sensible this Method of arguing is not new, yet the Observations here given may probably tend to illustrate the Case beyond what had been advanced on the Subject.

II. OPTICS

Logan's interest in optics arose in part from his studies in astronomy and astronomical instruments and in part from his love of working out mathematical problems. He was well aware of the phenomenon of atmospheric refraction and the illusory appearances created thereby,[1] and as the sequel will show, he was interested in refraction through glasses as a source of error in astronomical observations. His concern with the problem of accounting for error in the use of astronomical instruments was in the best tradition of theoretical science.

In June, 1737, upon perusing his article on the sexual generation of maize in the *Philosophical Transactions* of the Royal Society of London,[2] Logan read an article by John Hadley on combining transparent

---

[50] Stephen Hales, *Statical Essays* (2 v., London, 1733) 2: p. 291. The reference is to "Appendix, Observation XI, Experiment III." Hales suggested that the crooked appearance of lightning is a chemical phenomenon.

[51] "The Sentiments of the Reverend and Learned Dr. John Wallis R. S. upon the aforesaid Appearance, communicated in a Letter to the Publisher," Royal Society, *Philosophical Transactions*, No. 187: p. 323. Wallis's letter immediately follows an article on the same subject, written by William Molyneux and dated March 10, 1686/7.

(Notes followed by [L.] are Logan's. All other notes are the editor's.)

[1] See pp. 15 *supra*.

[2] See p. 38 *infra*.

lenses with reflecting planes.[3] The subject matter of the article inspired Logan to study Huygens's *Dioptrics*, quoted by Hadley, and he was astonished at what seemed to him the obscurity and tediousness of the demonstrations Huygens employed to explain his rules. Logan, therefore, attempted to develop simpler analytic and geometric demonstrations, and wrote them down first in English, as "The principal Rules in Dioptrics for finding the Foci of Glasses Demonstrated both by Analysis, and by Geometrical Construction." Logan translated this paper into Latin, and submitted the Latin copy to William Jones in the hope that it might be published in the *Philosophical Transactions*. If the Royal Society should reject the paper, it was to be delivered to Peter Collinson who would forward it to a friend in Holland, probably John Frederick Gronovius. Logan would have preferred that the paper be first published in England, although Holland would be an appropriate country for publication, both because that was Huygens's native land and because the Dutch had been soliciting Latin pieces from Logan for publication among them. Why Logan's paper was not published by the Royal Society is not clear, although Halley's intense dislike of the Pennsylvania Quaker may have been the determining factor. At any rate, Logan's paper was published at Leyden in 1739 as *Canonum pro inveniendis refractionum, tum simplicium, tum in lentibus duplicium focis, demonstrationes geometricae.* The English version is preserved in manuscript at the Historical Society of Pennsylvania. It is the English version, "The Principal Rules in Dioptrics," which appears below.

In the *Canonum* Logan conceded that Huygens's rules for the foci in dioptrics were "truly admirable" for their simplicity and universality of application. He complained, however, that, although the rules were "plain and easie," Huygens's demonstrations of them were too intricate and complex. He was particularly critical of the demonstrations in Proposition XII of *Dioptrica*, Book I, which he believed to be too long and perplexing to the reader.

The *Canonum*, on the whole, is an analysis of Huygens's propositions in *Dioptrica*, Book I, with particular emphasis on Huygens's rules in Propositions XII, XVI, XVII, and XX. It also contains a study of Propositions VII to XI, although Huygens had given no special rules in those propositions. It will be noted that Logan was chiefly interested in Huygens's rules and his demonstrations of those rules. What Logan tried to do was improve upon Huygens's demonstrations.

Logan began the *Canonum* with simple geometrical demonstrations of the ratio of the angle of incidence to the angle of refraction. This was elementary dioptrics, and the demonstrations were based on the generally agreed upon ratio of $I:R = 4:3$ between air and water. The purpose of Logan's figure 1, and his explanation of it, was to prepare the reader for his discussion of Huygens's propositions on refraction.

Logan dismissed Huygens's first seven propositions with one sentence: "The author in his first Seven Propositions, considers this, only as arising from plane Surfaces, which require no other Rule than the observation of the first principle."[4] Logan then proceeded to a consideration of Propositions VIII to XI.

Huygens, in Propositions VIII to XI, worked out problems in which he endeavored to find the point of dispersion of refracted rays in certain given conditions, i.e., a spherical convex surface of a transparent body on which parallel rays fall from outside; a spherical convex surface of a transparent body on which parallel rays fall from the interior; a spherical concave surface of a transparent body on which parallel rays fall from outside, and from the interior. Huygens concluded the problems worked out in Propositions VIII to XI with a definition: "Nous dirons que les rayons incident ou les rayons refractes correspondent au point vers lequel ils se dirigent ou dont ils proviennent ou semblent provenir."

Logan, in his consideration of Huygens's Propositions VIII to XI, attempted to show by fundamental demonstration "all the varieties there can be of parallel Rayes Suffering one Refraction only. . . ." He thought it necessary to go into more detail on these propositions than Huygens had so as to prepare the reader for his discussion of Huygens's Proposition XII.

Logan accepted the rules given by Huygens in his Proposition XII, but deplored Huygens's demonstra-

---

[3] John Hadley, "A Proposition relating to the Combination of Transparent Lens's with Reflecting Planes," Royal Society, *Philosophical Transactions*, Num. 440: p. 185. In an earlier article (Num. 420), Hadley had proposed the use of a telescope "with the Instrument for taking Angles. . . ." He subsequently considered combining a telescope with reflecting planes, and concluded, on the basis of Huygens's rules of dioptrics, that such an instrument would be subject to error, both because of "the spherical Figure of the Lens's, and also the different Refrangibility of the Rays of Light, when the Object is seen at a Distance from the Axis of the Telescope. . . ."

[4] The Dutch scholars who edited the works of Christiaan Huygens, published in 1916, wrote of his first three propositions in *Dioptrica*, Book I, Part I: "Ce qui nous frappe avant tout dans cette partie de la Dioptrique, c'est la grande rigueur que Huygens s'est imposée." ("What strikes us above all in that part of the *Dioptrica* is the great rigour which Huygens imposed.") "Avertissement-Premier Partie-Livre Premiere," *Oeuvres Complètes de Christiaan Huygens Publiées par la Société Hollandaise des Sciences*, **13**, 1 (La Haye, Martinus Nijhoff, 1916), p. xv. In his first three propositions and his definitions of the points of concourse or dispersion Huygens prepared his attack on the two principal problems he intended to solve in Book I: determination of the focuses of lenses and of places where are formed the images of objects whose position is given. *Ibid.* In Propositions IV to VII Huygens concerned himself with the location of the image in the case of a refracting plane surface, the rays being supposed to form small angles with the perpendicular on the surface. *Ibid.*, p. xvi. Logan gave more attention to Propositions VIII to XI than did the Dutch scholars in the 1916 edition.

tions of the rules,[5] which in this proposition filled thirty pages of Logan's copy of the *Dioptrica* and Logan believed them so complex that they could only confuse the reader. He, therefore, attempted to simplify Huygens's demonstrations by geometrical construction, and in far less space. The extent to which Logan clarified Huygens is a question which the historian must refer to the specialist in geometrical optics. However, the Dutch scholars who edited the 1916 compilation of Huygens's works pointed out that, after having determined in Propositions VIII–XI the focuses for a single spherical surface, he considered in Proposition XII "la situation de l'image d'un point lumineux produite par une telle surface. En traitant de ces questions il se sert d'un mode d'expression . . . qui lui permet d'énoncer les théorèmes d'un manière tout-à-fait générale et d'embrasser tous les cas des points lumineux et des images, réels et virtuels."[6] The variety of cases covered by Huygens's general theorems accounts for the length and complexity of his demonstrations. Logan, in his attempt to prove Huygens's theorems by geometrical construction, may have oversimplified for the sake of brevity and clarity, although he was fully aware of the comprehensive value of the theorems.

Logan passed over Huygens's Propositions XIII, XIV, and XV with a one-sentence description. Those propositions had no special rules given for them, and for that reason Logan thought it unnecessary to discuss them in detail. Propositions XVI and XVII involved a special rule for finding the foci or points of dispersion of parallel rays on convex and concave lenses which have unequal curvatures. Here Logan proved Huygens's rule by geometrical construction and algebraic analogy in several kinds of cases.

Dismissing Huygens's Propositions XVIII and XIX as unimportant in this context, Logan proceeded immediately to Proposition XX. He thought Proposition XX admirable, because it gave "one general rule for finding the foci or points of dispersion, of Rayes

flowing from or to, any point given in the Axis, for all manner of Lenses. . . ." He briefly explained the rule, and, employing the fifth figure in Proposition XII, demonstrated and proved the rule by analysis. Logan's appreciation of Huygens's rules in Propositions XII and XX is reflected in the comment by the Dutch scholars who edited Huygens's works in 1916: "Cette regle [in Prop. XX], ainsi que celle qu'il a donnée dans la Prop. XII . . . pour l'image produite par une seule surface réfringente, est présentée sous une forme très remarquable qu'on ne retrouve guère dans les traité's modernes."[7]

Logan fully accepted Huygens's rules in Book I of the *Dioptrica*. His only criticism was of the demonstrations of the rules, and he sought only to simplify and clarify these by analysis and geometrical construction.

After completing "The principal Rules," Logan studied the problem of spherical aberration in an effort to improve upon the work Newton and Huygens had done on this subject. Newton, in his *Optics* (1704), had offered a proposition in which he solved the problem of determining the focus of parallel rays which fall farther from the axis on a spherical glass and are refracted thence. He solved this problem through an infinite series evolved from surds.[8] Huygens' calculations were also limited to parallel rays, and, as Logan pointed out, "answers no further than the sines and versed sines hold in a duplicate ratio. . . ."[9] Logan's principal criticism of both Newton's and Huygens' calculations was that they ignored the "investigations of errors due to obliques, either divergent or convergent, though the knowledge of these is of even greater importance in Dioptrics than that of the right angle or the parallel ones." In telescopes "the rays fall obliquely on more surfaces of the lenses than at right angles"; in microscopes the rays always fall obliquely on the surfaces of the lenses. Newton had asserted that errors due to divergent or convergent rays could be resolved mathematically, but with greater difficulty. Logan developed a method which he believed to be "simple and clear," and he wrote that it extended "to all cases of oblique as well as parallel rays . . . and to all distances from the vertex, expressing in them all the quantity of the aberration in terms of the versed sines only. . . ." It appeared to him that the method he had worked out would account for the entire phenomenon of refraction in the science of dioptrics. The result of this study was another Latin paper, *De Radiorum Lucis in Superficies Sphaericas Remotius ab Axe Incidentium a Primario Foco Aberrationibus*. This work

---

[5] Huygens's general theorem in Proposition XII is as follows: "Etant donnée la surface spherique convexe ou concave d'un corps transparent et le point auquel correspondent les rayons qui tombent sur cette surface; construisons sur l'axe, qui passe par le centre et par le point donné, une quatrième proportionnelle à trois longueurs ayant chacune une extremité en ce point. La première de ces longueurs est la distance du point donné au point auquel correspondraient les rayons refractés provenant de rayons paralleles a l'axe venant de l'autre cote. La seconde est la distance du point a la surface refringente. La troisieme est la distance au centre de cette surface. La quatrieme s'étendra alors jusqu'au point qui correspond aux rayons refractés. Cette quatrieme distance doit etre prise a partir du point donné dans un sens tel que toutes les quatre soient dans le meme sens ou bien deux dans un sens, deux dans l'autre." Huygens divided this general theorem into eight parts to account for the variety of cases of rays passing through convex or concave lenses from the outside or from the inside and emanating from a given point or toward a given point. *Ibid.*, pp. 41–42.

[6] *Ibid.*, p. xvi.

---

[7] *Ibid.*

[8] The surd is an irrational root, e.g. $\sqrt{3}$.

[9] For example, if $a:b = b:c$, then $a:c = a^2:b^2$, and we say that a and c have the duplicate ratio of a and b. See Euclid, *Elements* V, Def. 9.

was published at Leyden in 1741.[10] Both the Latin original and an English translation appear below.

It would be presumptuous for the historian, who is not a professional researcher in optics, to evaluate Logan's work. As a matter of fact, few physicists who work in twentieth-century optics know very much about the history of their field. Nevertheless, it seems permissible to suggest, at least, that Logan may not have fully appreciated the extent to which Huygens was attempting to create a physical basis for his optics, rather than simply think in analytical or mathematical terms. Logan's demonstration on the basis of elementary geometry and trigonometry in the *Canonum* appears to be quite clear and convincing, but it seems unlikely that Logan made any significant contribution to the subject.[11] Of greater potential significance was his paper on spherical aberration. Here he was attempting to resolve a problem in optics in rigorous mathematical terms, although, as we are told, Newton and Huygens had believed it impossible. Logan's paper, brave though it was, proved to be of little significance in the history of optics. However, it illustrates the efforts of a colonial virtuoso to make a contribution to scientific progress in Europe.

LOGAN TO WILLIAM JONES, STENTON IN PENSILVANIA, 16th Oct. 1738, RIGAUD 1: pp. 335–341.
Esteemed friend,

. . . .

After I was, in June last, happily released from a charge, that, by the unkindness of a neighbouring government, had given me no small uneasiness,[12] looking again into the Philosophical Transactions, No. 440, in which something of mine had been most erroneously printed, making me, at the end of page 194, and beginning of the next, speak utter nonsense, I cast my eye on the next preceding article but one, which is a proposition by J. Hadley[13] relating to the combination of transparent lenses with reflecting planes, which affording me some matter for reflection, I from thence turned to Hugen's Dioptrics there quoted, (a work I had by me many years, yet till then had never read three pages in it,) and in viewing his twelfth proposition besides some others, I was shocked, I confess, at the tediousness as well as the obscurity of the demonstrations, (if they are to be called such,) that such admirable rules were attended with. Being therefore then at leisure, I resolved to try whether by some means or other I could not hit

on some plainer, and succeeded in it, as I have said in the introduction of the enclosed, so much to my satisfaction, both in the analytic and geometric way, and so far exceeding any thing I could meet with in the authors I have on the subject, that thinking it a pity the discoveries should die in my hands and be lost, I resolved to commit both methods to writing, which I did in English; but observing it was somewhat long, that Hugenius's work was in Latin, and that he himself had always in his writings preferred the methods of the ancients in demonstrating by geometrical construction, I chose to run this over again in that language, which is not unfamiliar to me, and send it thee in the manner it appears in the enclosed Latin copy, done by a pretty exact hand. Whether the Society may think fit to publish it in the Transactions, (for I have no way of preserving it but in some such collections, at home or abroad,) is left to your own judgment to determine. I need say nothing to recommend it; for the justness of the construction, the clearness of the demonstration, and the beautiful uniformity of the schemes, as applied to the several differences in the various cases, will do it abundantly to any mathematical reader; all which is owing to the subject, not to me, for I claim no merit from it, being conscious of having no other than the chance of an inventor, and the labour of digesting. But as I had drawn up both parts, the analysis and construction, as I have said, together in one, and afterwards chose to throw out the first, as judging the latter preferable, yet that first, if I mistake not, may upon examination be found worth preserving, and therefore I have got another hand to copy that also, which I here likewise enclose with the other, for thy own use, or to be applied as thou shalt judge proper, for I commit that wholly to thy disposal.

When I had gone through these, I proceeded to consider also the aberration of rays from the right focus, when they fall on the glass at a distance from the vertex, which Hugens had laboriously calculated; and Sir Isaac Newton in his Optic Lectures, p. 132,[14] has also a proposition for the same, which he solves by an intricate infinite series. For this also I happened to light on a simple and clear method. And as that of Hugens answers (I think) no further than the sines and their versed sines hold in a duplicate ratio, and neither of them for other than parallel rays, mine extends to all cases of oblique as well as parallel rays, though not with equal simplicity, and to all distances from the vertex, expressing in them all the quantity of the aberration in the terms of the versed sines only; so that from these processes, I believe, a good account may be given of the whole business of refraction as applied to that useful and entertaining science of dioptrics.

---

[10] According to Tolles, Logan, in his paper, "sought to prove that the laws of spherical aberration could be worked out with absolute mathematical rigor in spite of Huygens's (and Isaac Newton's) belief that it was impossible." (Frederick B. Tolles, *James Logan and the Culture of Provincial America*, p. 205.)

[11] At least no evidence to the contrary has as yet come to light.

[12] See Frederick B. Tolles, *James Logan*, pp. 170–178.

[13] John Hadley, "A Proposition relating to the Combination of Transparent Lens's with Reflecting Planes," January 9, 1734, Royal Society, *Philosophical Transactions*, Num. 440: p. 185.

[14] Probably Newton's *Optics*, Book I: Part I, Proposition 7, Theorem 6.

What I here now give, or have mentioned, may probably appear somewhat strange to thee from me, after my former honest declarations of the shortness of my skill in these sciences; but what I then professed I do sincerely still; for when I look into the works of divers others, I cannot esteem myself other than a smatterer in them, and but a mean one. The only advantage I have is a disposition to see, as far as my optics will carry me, into my subject, which, with application when I can use it, has sometimes helped me to a tolerable view of what I was in quest of. But little it is we do or can know here. In nature, scarce any thing beyond what, in Pliny's sense, may be called the history of it; in morals we may much more, as these are ordained for our grand duty in life, and powerful lights are breaking out on that quarter; but for bare speculation only, number and measure appear to me to be the most adequate objects on earth for the human intellect, (though this is infinitely short of being adequate to the extent of those,) but in the operations on the ideas of these there is certainty, to which in most other points (for faith is not knowledge) moral excepted, we are to remain, I doubt, for ever strangers. For this reason it is that, after a life of business, now at full sixty-four, though within these two years I intended the contrary, I can sometimes find as agreeable an amusement, not in reading but in thinking on these subjects, as in any that occur to me, and what I here give are some of the fruits, which had their first rise from the accidental occasion I have mentioned.

I shall add that, if this be published, I desire it may be done with due care and with justice to me, for I was ill used in the other. Nor can I think it was quite right, though there was no injustice in it to publish only that part of my letter to your president, about the angular appearance of the darts of lightning, which gave barely the thought, unattended, and consequently unsupported, by that clear instance I gave of the very same appearance in a straight rod or line, viewed at some little distance the wavy glass of a window, which puts it in every one's power to judge of the probability of the solution, when they might not otherwise see into the reasonableness of it.

I designed to have sent thee my whole English tract, but the copyist was obliged to leave it unfinished. He had, however, gone through all the analytic part, and entered on the geometric; therefore cutting off what little he had done in this, because imperfect, I send only that first, which is all that can be necessary, since you have the other fully in the Latin.

If the Society think fit to publish the Latin one, I request it may be speedily resolved, and if declined, that the piece, the Latin one I mean, after thy perusal of it, may be delivered to my friend P. Collinson, who, at my desire, will forward it to a friend in Holland, where I know they will be glad to publish it, and

perhaps that may be the most proper place: since Hugen's works were printed there. But I should rather choose to see it first appear in England. The reason of my mentioning Holland is, because I have been much solicited, by some of the most learned there, to suffer some other Latin pieces, they have seen of mine, to be printed amongst them.

I know not whether it may be to any purpose to request any kind of answer from thee. I am, however, for thy one past favour,

thy much obliged friend,
J. Logan

P. S. If the Society should incline to publish this Latin piece, but think it refers too particularly to a printed book, when it should rather mention the rules only, and give the demonstrations of them, I shall not be against an alteration, if made by a good hand, skilful both in the language and science, and with judgment, provided that the rules be mentioned as Hugens's, and nothing of mine be left out besides those references.

I was in hopes to have got the other smaller part or appendix, demonstrating the aberrations, ready for this opportunity, but could not. I propose, however, to send it this fall; but this need not retard the publication of the other; for in my opinion they would stand altogether as well in two several Transactions as in the same. If any errors appear either in the figures or writing, that have been overlooked, pray please to correct them.

(If to be favoured with an answer) I beg to be informed in it with what view the Society caused that proposition of J. Hadley, which I have in my letter modestly said afforded me some matter of reflection, to be published in their Transactions. For it is beyond my reach, I own, to find out the design of it. There is a star, placed in the figure, sending its rays to the speculum, which surely will be allowed, if any, to come parallel; and it is well known a speculum must reflect them equally parallel: and surely no mortal will ever allow that a telescope is to be so placed as to receive parallel rays otherwise than as parallel as may be to the axis. What then can be meant by those angles? Pray condescend so far as to inform the ignorant, for I am entirely so.

J. L.

"THE PRINCIPAL RULES IN DIOPTRICS FOR FINDING THE FOCI OF GLASSES DEMONSTRATED BOTH BY ANALYSIS, AND BY GEOMETRICAL CONSTRUCTION. BY JAMES LOGAN AT PENSILVANIA," LOGAN PAPERS, Vol. II: pp. 14–18, HSP.

As it is to Our sight we principally owe our knowledge, with most of our comforts in Life; and Vision by its only organ, the Eye, depends wholly on the Refraction of Rays through the transparent Coats and

Humors: we may justly wonder whence it came, that the Ancients who carried their Inquiries so very far into other branches of Mathematical Science, happened as far as we can discover, wholly to neglect this. On Optics and Catoptrics or Vision by direct and reflected Rays, we have two Tracts extant in Greek under the name of Euclid, in the 4th Phanomenon of which last, there is one effect of refraction mentioned, but not one word more is said of it in the Book. Hugenius[15] sayes, Aristotle amongst his Problems has one concerning an Oar appearing broken or bent in the Water, but it is a Mistake, for there is no such Problem there. Plutarch indeed mentions this[16] assigning its cause to the greater density of the water,[17] but the whole passage taken together will evidently shew, he knew very little, if anything, of refraction. A Tract of Aristotle's or one under his name on optics was in being, it seems, in the 13th Century, & Seen by the Author of the Speculum Doctrinale[18] &c and his contemporary tho' somewhat younger, the famous Friar, Roger Bacon, in his book de Perspectivâ[19] often quotes Ptolemy's Optics, but both these it Seems are now lost to us; and were they extant, it is not at all probable we should find in either of them anything to the purpose on Refraction. The Arabians therefore appear to us to have been the first who considered it. Alhazen a writer of theirs in the 11th Century in his large Treatise of Optics, in 7 Books soon after translated into Latin and Since printed, having in the last of them largely treated on Refraction. And Vitellio a Polander of the same 13th Age in the last of a larger work in ten Books, did the same. Friar Bacon also in his mentioned work treated particularly of the Subject, and from one passage in it, W. Molyneux[20] Supposed he fully understood the combination of Glasses and their effects,

but had that Gent. seen (for he cited at 2d hand) and considered that whole Chapt[21] he would probably have Suspended his faith in regard to that passage, no less than he would to the Friar's Story in it of Jul. Caesar's taking a prospect of the British Cities and forces by Specula erected on the Gallic Shoar. About the same time Peckam Archbp. of Canterbury wrote a Small piece on Optics, but intituled de Perspectiva, printed at Cologn in a thin 4 to, the most regular on the Subject of any thing of the kind I believe done in that age. Yet tho' much about the same time, or very soon after, the use of Spectacles was found out in Italy, it does not appear that any one of all these writers knew any thing from Experience, of the combination of Glasses,[22] before the Tubus Batavus or Telescope with two Glasses, a Convex & a Concave, was about the year 1600 accidentally discovered in Holl[an]d on the report of w[hi]ch Galileo invented his about 1608, published his Nuncius Siderius in 1610, and the very same year, the Sagacious Kepler wrote the first Piece on the Subject of Dioptrics that ever appeared, w[hi]ch he dedicated to his Patron the first of the following year. In this, that great author Supposed the Angles of Incidence and Refraction from Air to Glass to hold in all cases the same constant ratio of 3 to 2. Willebr. Snellius[23] found the ratio held truly constant, but not in the measure of the Angles themselves, but in that of their Secants Complexity from whence Descartes easily inferred, that then it must hold directly in the Sines, and this he published as his own.

After Kepler divers wrote on the Subject, with some further Improvem[en]ts, but they generally trod very much in the same path, till Dr Barrow's excellent Optic Lectures were published in 1669.[24] These

---

[15] Dioptr. pa. 1 [L.] The reference is to Christiaan Huygens. Notes followed by [L.] are Logan's.

[16] De Placitis Philosophorum Lib. 3, cap. 5 de Iside. [L.]

[17] διὰ τὴν πυκνοτέραν τοῦ ὕδατοσ ὕλην [L.] "... for the sight is forcefully bent aside because of the denser matter of the water." (Hermann Diels, ed., *Doxographi Graeci* [Berlin, 1958], p. 372. I owe the English translation of this passage to the Classics Department at Swarthmore College.

[18] Vincentius Bellovacensio/cited for this by Ger. Vossius De Scientiis Mathem. Cap. 26. 35. [L.] Gerardi Ioannis Vossii, *de Quator Artibus Popularibus, de Philologia et Scientis Mathematicus, cui Operi subjungitur, chronologia Mathematicorum. Libri tres.* (Amstelaedame, Ex Typographeio Ioannis Blaev, 1650). The second half of this volume is *De Mathematicarum Scientiarum Natura, ac Constitutione.* Cap. 26 contains the following: "Coeterum Opticum Aristotelis viderat quoque Vincentius Bellovacensis. . . ."

[19] Printed at Frankfort 1614. 4to in which pa. 84 he quotes Ptolemy's words from his 2d book, and de Speculis Mathem. pa. 37 he cites his 5th. [L.] An English translation of *de Perspectiva* may be found in Roger Bacon, *The Opus Majus of Roger Bacon* (2 v., Philadelphia, University of Pennsylvania Press, 1928), 2: pp. 419–582.

[20] Dioptr. pa. 274. [L.] Vitellio (or Witelo), a Polish philosopher in the thirteenth century, illustrated Alhazen's Optics in a treatise, *Optica*, published in 1270. (Joseph Priestley, *The History and Present State of Discoveries relating to Vision,*

Light, and Colours [London, Printed for J. Johnson, 1772], pp. 20–21.) Alhazen was an Arabian physicist, who was born about 965 A.D. and died about 1039. His *Optica* greatly influenced later European science.

[21] It is his last de Perspective & begins pa. 161: the passage mentioned is in pa. 167. [L.] William Molyneux (1656–1698) was the author of *Dioptrica nova, a Treatise of dioptrics in II parts* (London, 1692, 1709).

[22] Jo. Baptista Porta in his Magia Naturalis said to be first printed in 1589 has in Lib. XVII cap. 10 very plain hints concerning the combination of Glasses & their effects, yet 'tis doubtful whether he understood them for he advances many things in that book of Specula that are certainly false and groundless. [L.] Giambattista della Porta (1540–1615); his principal work was *Magiae naturalis sive de miraculis rerum naturalium*, lib. XX (Naples, 1589). In the foregoing Logan refers to John Peckham (d. 1292), *Perspectivae Communis Libri Tres, Iam postremo correcti ac figuris illustrati* (Coloniae, In officina Birckmannica, sumptibus Arnoldi Mylij, Anno 1592).

[23] Willebrord Snell, a Dutch mathematician, was born in Leyden in 1591 and died there October 30, 1626. Snell discovered the law of refraction during the later part of his life. His manuscript, which Huygens still could consult, is now lost.

[24] Isaac Barrow (1630–1677), *Lectiones opticae et geometricae: in quibus phaenomenon opticorum genuinae rationes investigantur, ac exponuntur: et generalia curvarum linearum symptomata declarantur.* Another edition was printed in London by G. Godbid in 1674.

Struck out new Lights to the whole Science of Optics, and in Dioptrics particularly, he gave or pointed out certain analogies, w[hi]ch had not been done before, for finding the foci, but it was generally by compounded Ratios. On these Dr. Gregory founded his, in his elegant little Tract. De Chales in his Cursus first published in 1674, but especially in the 2d posthumous Edition 1690 of that work, gave some very good Rules by Proportion.[25] W Molyneux also wrote well in the same way, but clogg'd his work too much with Calculations. But the illustrious Chr. Huygens Lord of Zulichem in Holland, applying his known great abilities to the improvem[en]t of Glasses and of this Science of Dioptics, in which he laboured many years, formed a much more compleat System of them than ever had appeared, (for Dr. Barrow & Sr I Newton gave only Lectures) yet he had not p[er]fectly finished it when he died in 1695, nor was it published by his Exec[uto]rs till 8 years after with his other posthumous pieces. In this the great Author gave Rules for the Foci, which for their Simplicity & Universality are truly admirable, & tho' they were not all of his own invention, yet by his accurate way of Applying them, he has in a manner made them all his own. But however plain and easie his Rules are, nothing of the kind perhaps was ever published more intricate and perplexed than his demonstrations of them, if they will bear that name. No less than 30 pages in 4to. has he bestow'd on one of them (Propos. XII) for Single Refraction only through its cases, and on one of these Cases (the 2d) he has above half a hundred comparisons of Ratios, many of them compounded, and very often in respect to their Majority or Minority, the most troublesome method of demonstrating, that the ancients (whom the author affected to imitate) have left to perplex their Readers, for tis with good reason believed of them, and we may be assured of the same in this author, that they seldom chose to demonstrate by the same means they invented. Shock'd at this, as I was of late occasionally considering one of his Propositions I resolved to try, whether much Shorter work might not be made of it by Analysis and I Succeeded in it with them all, even beyond expectation. I then further tried what then might be done by Geometrical Construction, and in this, by I know not what lucky chance, the Success was yet greater, for however easie & universal the Rules are, the Construction is not less so, but without recommending it, let the Readers to whom I shall here briefly communicate both methods, judge for themselves: But these Readers I must Suppose previously acquainted with the common Elements of Dioptrics, as the known Laws of Refraction, and particularly that I:R Stands

for the Ratio constantly observed between the Sines of the Angles of Incidence & Refraction, whatever the Ratio be, whether 3:2 between Air & Glass, which Huygens himself as well as most other authors generally take up with, or 5:3 the numbers given by Dr. Barrow, or 14:9, 17:11, 31:20, or 300:193, which on Several Occasions have been mentioned by Sr. Is. Newton. But the Ratio of 4:3 between air & water is generally agreed on. To render the whole more clear, I must premise the following as a lemma. If two

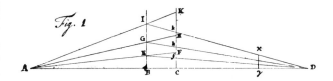

right lines BI, CK be raised perpend[icular] to a given Line AD, and from a point A in this, divers lines be drawn at different angles with the first AD, intersecting both these Parallels, as AF, AH, AK, and the Several distances of the points of Intersection in this last, from the base at C, as CK, CH, CF be divided in one constant ratio as of $I:R$ in k, h, f, respectively; If right lines be drawn from the points of Intersection in the first perpend[icula]r through these points of the ratio as Ik, Gh, Ef, and continued, they will either all meet in one point in the base on the side CD, or else on the Side BA on the other Side of A or before it, or otherwise they will all goe parallel between themselves & to the base BD Demonstr. By 2.6.El.[26] $AI: AK:: AG: AH:: AE: AF:: AB: AC$. So $BI: CK:: BG: CH:: BE: CF$.[27] Also Since by the hypoth[esis] $CK: Ck:: CH: Ch$ &c. It is $IG: kh:: GE: hf:: EB: fc$. Suppose CK in motion parallel to it Self tow[ar]ds D, the points k, h, f, c as they converge must always keep the same ratio in their mutual distances as at xy, till this line intirely Vanish and then all the converging lines must meet in one point as D.

Q. E. D.

But if CK remain greater than BI, tis plain the lines must then diverge towards the Side CD, and in the same manner with the preceeding they will converge towards the Side of A and terminate in a point beyond it, or further than A from B[.]

If it happen that $CK = BI$, it will necessarily follow from the parities of the Ratio before noted, that Ch must be $= BG$ & $CF = BE$ & consequently

[25] Claude François Milliet de Chales (1621–1678), *Cursus seu mundus mathematicus . . . Editio altera ex manuscriptis authoris aucta & emendata, opera & studio R. P. Amati Varcin . . .* (Lugduni, apud Anissonios, J. Posuel & C. Rigaud, 1690).

[26] Logan's reference is to Euclid, *Elements*, Book VI, Proposition 2.

[27] The proportions stated by Logan are read, for example, AI is to AK as AG is to AH ($AI: AK:: AG: AH$), but this example may also be written $AI: AK = AG: AH$, or $\frac{AI}{AK} = \frac{AG}{AH}$. The editor prefers to use Logan's symbols throughout.

lines through these points must be parallel to each other & all of them to *BD*.

Q. E. D.

III   Hence to proceed to Refraction. The author in his first Seven Propositions, considers this, only as arising from plane Surfaces, which require no other Rule than the observation of the first principle. In the next four the 8th, 9th, 10th, & 11th, He shews how to find the Foci or the points of concourse of Rayes moving parallel to the axis of a Spherical Surface dividing mediums of different Densities. And tho' in these there are no Special Rules given, yet in regard to what is to follow, tis necessary to Consider them.

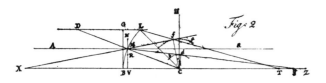

Let *BML* be a convex Spherical Surface of a body of Glass, the Center of its Sphericity *C*, its Axis *BZ*, and *A M* a Ray of Light flowing parallel to the axis in Air, and falling on the Surface at *M*. The angle of Incidence is *AMD* = *BCM*, the Sine of which is *MV*. Now this Ray entring a denser Medium must be refracted tow[ar]ds the perpend[icula]r *MC*, and the Sine of its refracted angle *CMS* must be to the Sine of Incidence *MV* in the Ratio of *R*:*I*, divide *VM* at *R* in that ratio, and with *Cd* = *VR* from the Center *C* draw the small arch *bdc*: by this arch touching it at *d* draw *MdS*, and *S* will be the focus of this Ray *AM*.

IV   Again Suppose the Ray *aM* moving parallel the contrary way and passing into Air, at *M*, Here as it is entring into a rarer medium, it will recede farther from *CM* the perpendicular to the Curve at *M*, and therefore will make the sine of its refraction greater than that of its Incidence. Accordingly take *VN*: *VM*:: *R*:*I*. Supposing *R* the greater term of the Ratio, and with *Cf* = *VN* from the Center *C* draw the Arch *fg*: from *f* touching this Arch, through *M* draw *f MX*, & *X* will be the focus of the Ray *aM*. But Since this is the Law of Refraction established in nature, that the Sines of the Angles of Incidence and Refraction Shall always be in the same Ratio; and in Triangles, it is an Axiom, that the sides Subtending the opposite angles are as the sines of those angles; In the △ *CRS*, *RS* will be as the Sine of the ∠ *RCS*, or which is the same, of *RCV*, the Angle of Incidence, and *CS*, as the Sine of the refracted ∠ *CRS*: So in the △ *CMX*, *XM* is as the Sine of *MCB* the ∠ of Incid[ence] and *CX* as the Sine of ∠ *CMX*, or which is the Same, of *CMf*, the refracted angle, w[hi]ch in this case, is greater than the first, that is *R*>*I*. So in the Triangle *LCT* from the Ray *DL*: *LT*: *CT*:: must be *I*:*R*:: *I* being the greater.

V   Now if we take *I*: *R*:: 3:2, or *v*:*v*. that we may the more clearly compare these lines, then *MS*: *CS*:: and *LT*: *CT*:: on the right hand must be precisely in the ratio of 3:2 and *CX*:*XM*:: the same on the other: but since in all these cases the Radius *CM* or *CL* etc., which is invariable, makes one of the sides of the Triangle, it is impossible that any constant ratio can be kept between the other 2 Sides, on *M* changing its place, the line *MS* always Shortening, as that recedes from *B*, but that the focal point must always change & draw nearer to the Vertex.   Thus *Z* is the focus for Rayes very near to the Vertex *B*, *S* of the Ray *AM*, and *T* of the Ray *DL*, and hence we may clearly see that such curve Surfaces can have no one certain general focus, but that which is taken for it, is that point only, in which the Rayes next to the Vertical point concur with the axis (for those that pass through that point itself are not refracted at all) and that is when both the Sine *MV* and the arch *MB*, become to all Sence the Same & coinciding also with the tang[en]t *BG*, as the Sines *Cd* & *Cf* will also at the same time with the perpend[icula]r *CH*. Therefore in all the following Propositions of the authors in which he gives rules for finding the principal focus, we need no otherwise consider the arch or its Sines, than as lines raised perpendicular to the axis taking the Incidence at any distance from the Axis, as in the Lemma, Thus.

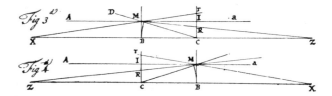

VI   As before let *BM* be supposed the tang[en]t of an outward convex Surface of a denser medium lying on the side of *a* and a Ray *AM* entring it at *M*, where notwithstanding its distance from *B*, the tang[en]t & arch are Supposed the same, Let *C* as before be the Center, from which the perpendicular *CIr* is raised. Now if *CI* be divided at *R* in the ratio of the Refraction, & *MRZ* be drawn, this will be the refracted ray, and *Z* the focus or point of concourse, and Since by the Triangle *BM* (= *CI*): *CR*:: *BZ*: *CZ*:: *I*:*R*. If we take *I* − *R* = *BC* the Rad. and accordingly lay off *BZ* = *I*, or *CZ* = *R*, we have the focus given.

VII   In the same manner also as before, if the Ray *aM* pass out of a denser into a rarer medium through a Surface convex to the rarer, that is through the same *BM*: Seeing it will be refracted from the perpend[icula]r *CD*. If we take *Cr*: *CI*:: in the ratio of the Refraction and draw *rMX*, it is plain, as before that *CX*: *BX* are also in the ratio of the Refraction & *CX* = *I* or *BX* = *R*.

VIII   Again If a parallel Ray *AM* (fig. 4) from a rarer to denser medium fall on a Spheric Surface, whose Center is C, Concave tow[ar]ds the rarer, as it enters a denser it will incline to the perpend[icula]r CM and as this according to the course of the Ray diverges from the Axis, so will the Ray it Self and its focus will be what is called an imaginary one, or more properly a point of dispersion at Z, at the same distance from B as Z in the first above[.] So if the Ray moving the contrary way from a to M (fig. 4) from a denser to a rarer, it will recede further from MC the perpend[icula]r to the curve, and taking rC: IC:: in the Ratio of the Refraction it will pass from M to r, as if it flow'd from X its point of dispersion w[hi]ch is at the same distance from the Center C as X is in the 2d case of the next preceeding figure.

IX   Thus are all the varieties there can be of parallel Rayes Suffering one Refraction only, fundamentally demonstrated. We are next to prove to the author's XIIth Proposition, which is for finding the focus or point of Dispersion of Rayes proceeding from, or tending to a given point in the Axis, in rarer or denser diaphanous mediums, and refracted by either convex or concave Spheric Surfaces. For which in all its possible cases he gives this one general Rule.

These three, the vertex of the Surface, its Center C, and the mentioned point A being given, we are further to take for a 4th, R the point in the axis, w[hi]ch would be the point of concourse of parallel rayes refracted by the Same Surface & coming from the Opposite Side to A. Then however A be Situate, if to AR, AB & AC we take a 4th proport[iona]l AZ, in the same axis, Z will be the point Sought. But these four must either be taken all the same way, or two one way, and two the other.

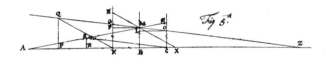

X   Let BM be Supposed the Tang[en]t of a Spherical Convex Glass, the Radius of which is BC, its Center C, its Axis ABC, AM a Ray from the given point A falling on the Surface M Supposed very near to B: Continue AM till it intersects the perpend[icula]r CN raised on the Center C in N: Divide CN at O in the ratio of the refraction: through O from M draw MOZ, and Z will be the focus, or point Sought, on which to find and demonstrate the Rule. Draw LO parallel to BC, then $BL = OC = (R/I)\ CN$, & $LM = BM - (R/I)\ CN$ and the Triangles MLO, MBZ are Similar, therefore: $BM - (R/I)\ CN$: $(LO =)\ BC$:: BM: BZ. But AB and BC being given, BM: CN:: AB: AC. Substitute these latter in the place of the first, and we shall have $AB - (R/I)AC$: BC:: AB: AZ. Put $AB = a$, $BC(=\text{Rad.}) = r$, $I = m$, $R = n$, then $a - n/m$: $a + r$: r:: a: $z = am$ $- an - nr$: mr:: a: $z = \overline{m - n}a - nr$: mr:: a: z.

Lay off in the fig. CR: BR:: I: R:: m:n, and R will be the focal point for parallel Rayes mentioned in the Rule, and $m - n$ will be $= BC = r$, whence we shall have for the preceeding Analogy, $ar - nr$: mr:: a: z and dividing the first by r, $a - n$: m:: a: z; whence $z = am/a - n$ then by composition $a - n$: $a - n + m$:: a: $a + z$ but $a - n + m = $ (as before) $a + r$, theref[ore] tis $a - n$: $a + r$:: a: $a + z$ or conversely $a - n$: a:: $a + r$: $a + z$ which is exactly the Rule that was to be Demonstrated.

XI   For another example, Let us Suppose the Ray GM going out of a convex Surface of Glass BM, whose Center is K, in the direction MZ to find the point to which on its egress it will be refracted. Take, as in the like Case in the preceding, for parallel rayes KH:KG in the ratio of the refraction from H through M draw HMX, and X will be the focus Sought. For the Rule, Draw FM parallel to KB, and FH will be $= (I/R)\ KG - BM$ [.] The Triangles FHM and BMX are similar: theref[ore] $(I/R)KG - BM$: $FM\ (= KB)$:: BM: BX: But, as before KG: BM:: KZ:BZ. Substitute these latter, and we have $(I/R)ZK - ZB$: KB:: ZB: BX.

Put $ZB = b$, $KB = r$, $ZK = b + r$, $I = m$, $R = n$, $BX = x$, then $(m/n)\overline{b + r} - b$:$r$::$b$: $x = mb + mr$ $- nb$: nr:: b: $x = \overline{m - n}\ b + mr$: nr:: b: $x = br$ $+ mr$: nr:: b: $x = $ (as before) $b + m$: n:: b: x. theref[ore] $x = nb/b + m$. But further by Division of proportion, $b + m$: $b + m - n$:: b: $b - x$ that is again as $b + m$:$b + r$::b:$b - x$. Lay off again from B. BP: KP:: I: R:: m: n, then $m = BP$, $n = KP$, $m - n$ (as before) $= KB$ whence we have ZP: ZK:: ZB: ZX which is again exactly the Rule, that was to be demonstrated.

XII   Having after I had Struck out this method of Demonstration by Analysis discovered the other that is to follow by Geom[etrical] Construction, I so far preferr'd the latter, that I thought of taking no further notice of this first. But observing on further reflection, that this analytic method, not only fully demonstrated each Rule, but at the Same time also gave with it, all its proper distinctions, I could not but Judge it well worth our notice. For as before in finding the focal point for parallel Rayes, we Saw it fell in some Cases, at the distance of thrice the Radius from the Vertex, and in others of twice the Radius only. So here, in bringing out the Rule, it brings its direction with it, in that point also, as in the first Case we have $a - n$, in the other $b + m$—marking out to us in each, the focal distance for parallel Rayes, from the Vertex as well as giving us the Rule we Look for. And the like will agreeably occur in other demonstrations that follow. And the two following are added here for a particular Variety in them tho' they are both for Rayes passing out of a denser into a rarer medium by a Surface concave in the denser.

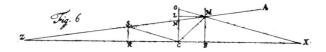

XIII   Let AM be Supposed a Ray passing out of a denser medium at M as the tang[en]t of a Spheric Surface concave tow[ar]ds its Center C, in the direction of MZ. To find the refraction of which, we are to lay off $CO: CN:: I:R$, and drawing OM, being continued it will give X the point of dispersion. And to apply the Rule, without repeating the former work, putting as before $ZB = b$, $CB = r$, $I = m$, $R = n$, $Bx = x$ the Similar Triangles OLM, MBX, give this Analogy $(m/n)t - r - b:r:: b:x = mb - mr - nb: nr::b:x = rb - mr:nr::b:x = b - m:n::b:x$ and by Composition, $b - m: b - m + n::b: b + x = b - m: b - r:b: b + x$, that is $ZR: ZC:: ZB: ZX$ which is the Rule QED.

XIV   Again let us Suppose the like Ray AM moving in all respects as the other, Save that Z to which it is directed, lies between R and B, as in the other it lay beyond it. Here the Similar Triangles MLO, MBX

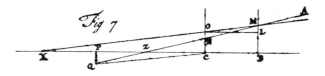

give this analogy, $b - (m/n)\overline{b - r}: r:: b: x = nb - mb + mr: nr:: b:x = mr - br: nr:: b:x = m - b: n: b:x$, and by Divis[ion] of Reason, $m - b: m - b - n (= r - b):: b:b - x = ZR: ZC:: ZB: ZX$ the Rule, QED. In this last analogy, we have $m - n - b = r - b$ which is a negative for $r < b$, which plainly points out to us, that we are to take ZC the contrary way from ZR, and $b - x$ is also just the same for $x = BX$, & $b = BZ$ theref[ore] here $b - x = x - b$ taken the other way. So in the next preceding Analogy, we had for the first term $ZR = b - m$, but here it is $ZR = m - b$, as the one happens to be greater than the other, and these are further Instances of what I had just hinted before.

XV   The Author's three next Propositions for finding the focus of a whole Sphere (which falls at about the distance of half the Radius beyond it) and of planoconvex, and plano-concave Lenses, both which have their focus at the distance of about twice the Radius from the Vertex, and very nearly equally, whether the Rayes fall on the plane or the curve Surface. These Propositions, I say, having no Special Rules given for them need not be insisted on here, & therefore we may proceed to the next that hath one.

XVI   The XVI and XVII Propositions are for finding the foci of all other kinds of Lenses for parallel Rayes. For Convexes or Concaves of equal Sphericities, there is but one easie Rule, which Shews that the Focus or point of dispersion, for the ratio of $3:2$, falls on the Center. But when the Curvatures are unequal, it must fall differently, for which, for convexes, and for a Meniscus the Cavity of which is of a greater Radius than the Convexity, the Author gives this Rule.

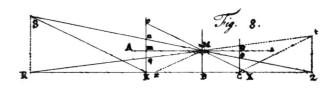

XVII   If BM be Supposed the tang[en]t of the Surfaces of a Glass convex on both sides the radius's of which are BC, BK, their Centers C & K (the thickness of the Glass not being now considered) for BC, take BZ, $CZ::$ & for BK, $BR: KR:: I:R$. The Rule then is $ZR: ZB:: ZK: ZX$ and X is the focal point.

Demonstr.   Put $BC = a$, $BK = s$ and as m and n are taken for the ratio of $I:R$, $BZ = am$, and $BR = sm$. By the Rule of the 16th Propos[ition] putting $BZ = b$, BX will be $x = bn/b + m$, that is $b + m: n:: b: x$. But here we are to observe, that m and n, being only the terms of a ratio, they have no determinate Value, in regard to quantity, but to proportion only: and $m - n$ always Signifies the Radius, or that line to which the ratio is applied, be it greater or less. We must therefore in this last Analogy for m, & n; Since they relate to another Rad. $= s$, take sm and sn, and the analogy will then be $b + sm: sn:: b: x$. For b Substitute its preceding value am, and it will give $am + sm: sn:: am: x$, and by division of Reason $am + sm: am + sm - sn:: am: am - x$. Now for the above mentioned reasons, $\overline{sm - sn} = m - \overline{ns} = s:$ therefore it is, $am + sm: am + s::am: am - x$, and Conversely $am + sm: am:: am + s: am - x$ that is $ZB + BR:ZB:: ZK: ZX$ which is the Rule Q.E.D.

XVIII   But if we take the preceding analogy, before division $am + sm: am:: sn: x$, and divide the first part by m, we shall have this much easier analogy $a + s: a:: sn: x$ that is $x = asn/a + s$ which we may resolve either into the preceding, or, into $a + s:s::an:x$ that is as the Sum of the two Radii is to one of them, so is n multiplied into the other, to the distance of the focus from the Vertex of the Glass. If the Ratio of the Ref[rac]tion be taken $3:2$, $n = 2$ if $14:9$, $n = 9/5$, if $4:3$, as between Air and Water $n = 3$. If $s = a = 1$, $asn/a + s = 2/2 + 1$ & $x = 1$ Rad.

XIX   For Lenses concave on both sides equally or unequally the Rule and its demonstrations are exactly the same, for their point of dispersion, w[hi]ch will always lie on the contrary Side. But for a Meniscus, the case is altogether different, as the Centers of both their Sphericities must lie the Same way, and they have two Varieties, the one when the convex Side is of a greater curvature, that is, has a less Radius than the Cavity, and the other when the Convex has the greatest Radius, when they are equal they are of no use in Dioptrics, for the Rayes go out at the 2d Surface parallel to themselves as they entred the first. And between this as one extream, and the Plano-convex, all the Varieties of the first kind lie; and between the same and a Plano-concave all the Varieties of a 2d ffor in the first the Cavity may so lessen as to to [sic] become a Plane; and as the Plano-convex Glass has it's [sic] focus at the distance of about twice the Radius before the Ray or the same way it tends to: the focal point of every such Meniscus must lie beyond that distance & may enlarge til 'tis lost in Parallelism. So in the 2d kind when the Convex Side has the greater Radius, that convexity may lessen till it become a Plane and the Lens become a plano-concave, w[hi]ch throws its point of dispersion to the distance of about twice the Radius behind it; and as that plane becomes more convex, the point of dispersion will still recede and enlarge, till lost like the other in parallelism, the contrary way; as in these figures here added, for the clearer explanation of these Glasses which are the most difficult of all others[.]

XX   The first of these is for Glass of equal Sphericities, the 2d a true plano-convex Lens, giving the focus at the distance of the Diam[ete]r nearly, the 3d is a Meniscus Somewhat concave throwing its focus to

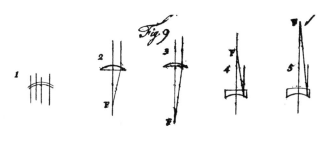

a greater distance, the 4th is a true Plano-concave w[i]th its point of dispersion at the distance of its Diam[ete]r the other way, the 5th a Meniscus with Some convexity throwing its point to a further distance, and the extreams of the first Meniscus fig. 3 we See are the 1 & 2d figg, and of the other fig. 5 the 1st & 4th.

XXI   These foci or points are found by the Same Rule in the Author's with the preceding but with this difference, that in Lenses convex or concave on both sides, the distance taken for the first term of the Ratio is the Sum of the Radii enlarged but in these it is their difference as in the figures, in the first of which fig. 10, BM is first Supposed the tang[en]t of a Convex Lens tow[ar]ds A, the Rad[ius] of w[hi]ch is BC, refracting parallel Rayes to Z. The Same BM is also to be Supposed the tang[en]t of a Concave tow[ar]ds

a, the Rad[ius] of w[hi]ch is BK, refracting MZ as passing into Air into MX. In the other fig. 11, BM is first Supposed the tang[en]t of a Convex whose Rad[iu]s is the greater BC, and again of the Concave of a lesser Rad[ius] BK, refracting MZ from the Axis, by taking $KO: KN:: 1:R$, So far as to render it diverging & as taking its course from X, the other way in the Axis. And to demonstrate the Rule of Analysis for both these, we are only to Observe, that Z in these figures alwayes pointing out the focus for parallel Rayes falling on the first Surface, and R denoting also what would be the focus of parallel Rayes refracted by the 2d Surface.

XXII   When Z lies between the Glass & R as in fig. 10, we are to find X, as in the work relating to the preceding fig. 7 §XIV but when Z lies more remote than R, we are to take the work of fig. 6, §XIII and as $BZ = b$, is in all these Cases of parallel Rayes = am, taking $a =$ Rad. of the first Surface, $S =$ to Rad. of the 2d, and sm, sn, for m, n in this 2d We shall have for fig. 10, $sm - am: s - am:: am: am - $ x that is

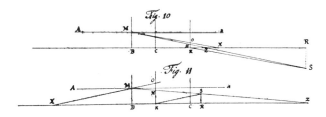

$ZR: ZK:: ZB: - ZX$ in which analogy, the 2d & 4th terms are both Negative, because they are to be taken the contrary way. For fig. 11, from the work of fig. 6, §XIII we have $am - sm: am - s:: am: am + x$, for here $a > s$, that is just the same as before $ZR: ZK::ZB: ZX$. In both which the Rule is fully demonstrated. But from the same work this Rule (as in convexes & concave) may be shortned by taking $x = asn/a \cdot s$ w[hi]ch may be Naturally deduced from the first of the work, as in the other, that is $a \infty s: a:: ns: x.$ or $a \cdot s: s:: na: x$, w[hi]ch is as the difference of the Radius's is to one of them, So is the other multiplied by n, to the distance of the focus or of the point of dispersion from the Glass.

The Authors two next Propositions being particular Problems, require not to be Spoke to here, & therefore we may proceed to the following.

XXIII   The next Proposition (the XXth) is an Admirable one, as it gives one general Rule for finding the foci or points of dispersion, of Rayes flowing from or to, any point given in the Axis, for all manner of Lenses, and tis briefly this.

XXIV   We are first to find the focus or point of dispersion of the Lens for Rayes parallel to the Axis. Then as the distance of the given point from the focal point is to its distance from the Glass; So is this last distance to the distance of the focal point Sought from the point given, and to demonstrate this most comprehensive Rule by Analysis we need only to return to the preceeding work of the XIIth Propos[ition] and take an example in the 5th fig. where AM is Supposed a Ray from the point A falling on the point M of BM Suppose the Tang[en]t of a Surface of Glass convex tow[ar]ds A whose Center is C, and it was there found that this Ray would by that Surface be refracted to Z. But in the same figure, the Same Ray GMZ was considered as passing out of Glass into Air by another Surface of Glass convex tow[ar]ds Z of w[hi]ch BM was also Supposed the Tang[en]t and K its Center, both Surfaces having their Radius equal as $KB = BC$, and this 2d refraction was found to be MX and consequently X the focus of the whole Lens, now to apply this to the Rule and to demonstrate it.

XXV   In the first part of the Analytic Operation on that Proposition XII & figure 5th §10 putting $AB = a$, BC (= Rad] = r, $I = m$, $R = n$ we found $BX = am/a - n$ and by the 2d part of the work, §XI putting the same $BZ = b$, we found $BX = nb/b + m$ that is $b + m$: $b::n$: $BX = x$. Now since $b = am/a - n$ let this be Substituted in the place of b, in the last Analogy, & it will give $am/a - n + m$: $am/a - n::$ n: $x = am + am - na$: $am::$ $n:x$, and dividing by m, $2a - n$: $a::$ n: x, conversely $2a - n$: $n::a:x$ by Composit[ion], $2a - n$: $2a - n + n::a$: $a + x = 2a - n$: $2a::a:a + x$ divide the first part by 2, and we have $a - n/2$: $a::a$: $a + x$, which is exactly the Rule Q.E.D. with this further direction, that what the Author in the rule calls the focus for parallel Rayes we here find must be $n/2$ which that focus for Such Lenses truly is, & therefore if the ratio between Glass and Air be $9:14::n:m$ that focus will not be $=$ Rad. $= 5$ but $= 4\frac{1}{2}$ and so it will be truly found as has been tried by Trigonometrical Calculation: For air to water the focus will be 3/4.

XXVI   When the Convexities are different, we are as in the last preceeding Cases for the Meniscus, to take the 2d m, n in the ratio of the 2d Diam[ete]r as s keeping a for the first as before. Thus for $b + m:n::b:x$ we are to take $b + sm$: $sn::$ $b:x$ and Substituting $am/a - n$ for b we have $am + asm - asm$: $am::$ $sn:x = a + as - ns$: a $::$ ns: $x =$ (by composit[ion] & convers[ion]) $a + as - ns$: $a + as - ns + ns$ $::$ a: $a + x = a + as - ns$: $a + as::$ a: $a + ax$. Say $a + as$: $a::$ ns: $ans/a + as = ns/s + 1$, Substitute these, that is a for $a + as$ & $ns/s + 1$ for ns, & we shall have $a - ns/s + 1$: $a::a:a + ax$, which again is exactly the Rule, for $ns/s + 1$ is the true focus of this Lens for parallel Rayes as we have Seen in §18 but are to remember that 1 and n both have relation to a only, and here again is a beautiful instance of the excellency of this method in its bringing out the Rule itself, together with its demonstration.

XXVII   To proceed in multiplying examples of which there is a considerable Variety would not only enlarge this too much, but might also Anticipate the pleasure a Reader may give himself in making his own Trials, it may therefore Suffice to Say, that the first refraction z or b will always be $am/a - n$ or $am/n - a$ and the 2d, or x will always be $= snb/b + sm$ or $snb/b \backsim sm$ and these two combind, as above, and the Analogy either compounded or divided, will always bring out the Rule, for it holds in all cases, and therefore all the Rules are by this Analysis clearly demonstrated.

The Same Rules demonstrated Geometrically by Construction

XXVIII   Refraction with its Rules in Dioptrics have been so fully explained in the preceeding Demonstrations by Analysis, that this Second method by Construction or Geometrical Effection, which tis presumed will be found not to yield in beauty to anything that has yet appeared in Geometry, may be much more briefly exhibited. For notwithstanding no Propositions can be more Universal, they are yet at the Same time equally Simple Clear & Easy. And as they proceed on the Same fundamental Scheme that has been apply'd in the preceeding, the same Schemes & figures will for the most part with some Small Additions equally Serve also for this method.

XXIX   The first of the Authors Propositions, there is any occasion to take notice of in this way, is his XIIth: for all the other preceeding it, when considered geometrically in the Schemes, demonstrate themselves. Let us theref[ore] begin with that XIIth, for finding the focus for Rayes issuing from or tending to a given point in the Axis, w[hi]ch has been fully demonstrated by Analysis §9, 10 &c.

DE RADIORUM LUCIS IN SUPERFICIES SPHAERICAS REMOTIUS AB AXE INCIDENTIUM A PRIMARIO FOCO ABERRATIONIBUS, LEYDEN, 1741.

I.

Cum utilissima simul & jucundissima haec scientia, quae Dioptrice dicitur, sit ex mixtis seu Physicomathematicis una; ideoque ea pars ejus, quae Physicis innititur rationibus, rigorem Geometricum in quibusdam vix admittat, eam vero in praestantissimo opere suo ita excoluerit vir Clarissimus Chris. Hugenius, ut omnibus omnino usibus sufficere videri possint, licet a rigore illo haud parum desciscant, quas tradidit, regulae. Is ideo operam ludere fortassè videbitur; qui in ulteriore earum scrutinio aliquam ponere velit. Ego vero, cum post eam, quam in canonibus ejus clarâ luce Geometricâ donandis collocaveram, illa etiam, quae de radiorum ab axe remotiorum per Lentes aberrationibus insuper tradidit Autor, penitius perpenderem, & unde & quo calulco [calculo] numeros suos absolutos extuderit, quos usui commendat, accuratius inquirerem, occurerunt mihi quaedam rem ipsa haud parum illustrantia, & quae secum communicari haud aegrè, opinor, feret harum Studiosus Lector.

II.

Unde autem oriantur hae, quas hic intelligo aberrationes, in partis hujus prioris §§4ta. & 5ta. abunde demonstratum est. Nimirum in fig. 2dâ. cum ipsius superficiei *BML* verus seu proprius focus sit punctum *Z*, radii vero *AM* ab axe remotioris sit *S*, & radii *OL* sit *T*, horum distantia a puncto *Z*, nempe *ZS*, *ZT* sunt ipsae, de quibus hic agitur, aberrationes.

III.

De his investigandis nullus, quod sciam Dioptricorum scriptor, praeter hunc unum, meminit, nedum quicquam tradidit, praeterquam quod in Clariss. Newtoni Lectionibus Opticis olim Cantabrigiae habitis, nuperius vero in lucem editis, occurrat propositio (Props. XXXI. Page 132) pro Radiorum parallelorum ab axe remotiorum in sphaeram incidentium, & inde refractorum, focis determinandis; quam, methodo satis quidem perplexâ, per seriem infinitam e surdis erutam solvit. Quae tamen series, quamvis sane egregia, ut ibi ab Autore traditur, duplici laborat defectu, ut ex sequentibus abundè patebit. Neuter

autem horum pro investigandis obliquorum, seu divergentium aut convergentium erroribus, quicquam tradidit: cum horum tamen scientia etiam majoris sit momenti in Dioptricis, quam rectorum sive parallelorum. In omni enim organo dioptrico composito, radii in plures lentium superficies incidunt obliqui, quam recti: in Telescopio duarum Lentium, si in primam objectivae (ut dicitur) incidant paralleli, in tres alteras impingunt obliqui: si quatuor sint Lentes, in duas incurrere possunt paralleli, at in omnes sex reliquas incurrent obliqui: & in Microscopiis in omnes obliqui. Newtonus quidem in Scholio ad Propos. praedictam pag. 136, ajit, eodem modo, quo errores parallelè incidentium determinavit, consimiles divergentium vel convergentium, licet calculo difficiliori determinari posse, de quibus postea.

IV.

Ipse autem Hugenius in Propos. XXVII. modum per surdos, eumque perplexum & difficilem, hos errores investigandi monstrat. Ut laboris vero taedio parcatur, numeros pro parallelis in omne genus lentium incidentibus tradit, de quibus in sequentibus etiam videbimus. Horum autem prioris, nempe posito sinu verso = v pro radiis è vitro in aëra exeuntibus, errorem dicit fore $(9/2)v$, & contrà ex aëre in vitrum $(4/3)v$: at hi numeri non ultrà comperientur veri, quam

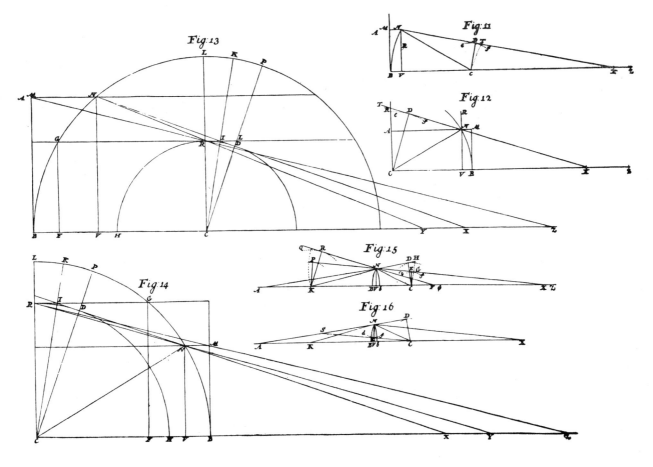

Fig: 13   Fig: 11   Fig: 12   Fig: 14   Fig: 15   Fig: 16

quousque inter sinus rectos & versos, obtinet ratio duplicata, quod primum aut alterum vix excedit gradum. Sed res tota hoc pacto, ni fallor, longe clarius & facilius expedietur.

Sit in fig. 11. & 12. BN superficies sphaerica convexa ad partes M, cujus centrum C, axis CB; sit AN radius lucis axi parallelus superficiei impingens ad N, in fig. 11mâ. ex aëre in medium densius; in fig. 12mâ ex hoc in aëra. Ex praeceptis in priore parte traditis, si hic radius impegisset quam proximè vertici B, focus ejus incidisset in axem ad Z: fuisset scilicet $ZB:ZC::I:R$. Nunc vero, sicut in partis prioris §VI. demonstratum est, ut ab isto foco deficiat necesse est. Sumptis igitur in NV sinu recto $VN:VR::I:R$, intervallo $CD = VR$ è centro C describatur arcus eDf, ad quem ab N ducatur tangens NDX, & verus focus hujus radii NX erit ad punctum X.

V.

Ad hunc focum determinandum per calculum; ponatur $BC(= NC)$ Radius $= a$, Sinus $VN = m$, $CD(= VR) = n$: & cum sint $VN: VR::I:R$, erunt etiam $I:R:: m:n$. Sit BV Sinus versus $= v$, unde erit $CV = a - v$; & cum habeantur sinus versi in tantillis arcubus, ad se invicem in ratione duplicatâ rectorum suorum, sinus versus recti CD erit $(n^2/m^2)v$; unde etiam ND erit $= a - (n^2/m^2)v$: sit porro $CX = x$. Triangula NVX, CDX, propter angulos rectos ad V & D, & communem ad X, sunt similia, unde $VN: VX::CD:DX$, h.e. in fig. 11mâ. $m:x + a - v::n:nx + na - nv/m = DX$; & hinc

$$n\frac{\overline{x + a - v}}{m} + a - \frac{n^2}{m^2}v = NX$$

sunt vero $NV: CD::NX:CX$. $h.e.m:n::$

$$\frac{n}{m}\overline{x + a - v} + a - \frac{n^2}{m^2}v:x,$$

unde aequatio

$$mx = \frac{n^2}{m}\overline{x + a - v} + na - \frac{n^3}{m^2}v,$$

quae ritè ordinata dabit $x = (mn + n^2/m^2 - n^2)a - [n^2 + (n^3/m)/m^2 - n^2]v$. Dividatur hujus prima pars per $m + n$; est enim $m^2 - n^2 = \overline{m - n} \times \overline{m - n}$; & in omnes partes 2dae. ducatur m, fiatque divisio per $m + n$, & erit $x = (n/m - n)a - (n^2/m^2 - mn)v$.

VI.

In fig. 12mâ., erit $VX = x - a + v$ (nempe positâ, ut prius, $CX = x$) & $NX = (n/m)\overline{x - a + v} - a + (n^2/m^2)v$, & si eodem modo in hâc pergamus, quo in superiore, & per omnia mutentur signa, ipsa eadem proveniet aequatio (posito n hic majorem esse rationis terminum) nempe $x = (n/n - m)a - (n^2/mn - m^2)v$. In quarum aequationum utrâvis sumptis, more

Hugeniano, $m:n = 3:2$ seu $2:3$, eosdem omnino, ac ille, habebimus numeros: scilicet in hac posteriore $3a - (9/2)v$, in priore autem $2a - (4/3)v$. Quinimò dant ipsissimos terminos seriei Newtonianae, de quibus mox erit dicendum. Primus terminus enim $(n/m - n)a$ est ipsa CZ; altera vero $n^2/mn - m^2v$, est ipsa aberratio, scilicet ZX, auferenda ex CZ; Resolvetur autem in has partes $n/m - n \times n/m$, quae pro valore m & n erunt vel $2/1 \times 2/3 = 4/3$, vel $3/1 \times 3/2 = 9/2$.

VII.

Sed in praecedente processu sumebatur pro concesso, sinus versos DT, in utrâque figurâ, esse in duplicatâ ratione sinuum suorum rectorum VN & VR seu CD; quod tamen ultra primum aut alterum Quadrantis gradum, ut dictum est, vix obtinet. Ideo quaerenda est regula universalior; quam ex iisdem figuris facillimè obtinebimus, si pro $(n^2/m^2)v$ substituamus $y = DT$, cujus valor hauriendus erit ex sinuum Tabulis, deducendo scilicet ex Radio ipsum ND, qui est sinus complementi ipsius CD, atque ita dabitur aequatio $mx = (n^2/m)\overline{x + a - v} + na - ny$, unde ex calculo praecedentis omnino consimili, habebitur $x = (n/m - n)a - n^2v + mny/m^2 - n^2$, cujus pars postrema, per omnes circuli gradus, in quibus fieri possit refractio, erit pro radiis parallelis, vera aberrationis quantitas. Resolvi autem poterit in hanc $\overline{nv + my} \times n/m^2 - n^2$ vel etiam in hanc $n/2 \times \overline{v + y/m - n + v} \infty y/m + n$. Utendum autem erit signo $-$ quum n est rationis pars minor; quum vero major, signo $+$.

VIII.

Sed ad majorem rei claritatem, dispiciamus unde conficiatur haec aberratio ZX, quod ex hâc constructione. Geometricâ non tantùm liquidò constabit, sed etiam ex eâdem, modus eam eruendi & determinandi planior, & forte aliquanto facilior exhibebitur. Sit itaque in fig. 13tâ. & 14tâ., AM radius lucis axi parallelus impingens in superficiem BNL convexam ad partes M, in fig. 13tâ. è medio rariore in densius, contrà in fig. 14tâ. Si hic radius quam proxime vertici B inciderit, sumptâ CR ad sinum $NV = MB$, in ratione refractionis, & ductâ MRX per punctum R in normali CL, daretur punctum Z (ex dudum in parte priore demonstratis) focus hujus superficiei primarius: impingit vero radius ad N: agatur igitur ab N per idem punctum R recta NRY, occurrens axi ad Y; unde propter parallelas BZ, AN, erit $ZY:MN::ZR: MR::YR:NR$, hoc est in fig. 13tâ. (sumptâ Hugenii ratione $I:R::3:2$) YZ erit dupla sinus versi MN; at in 14tâ., ejus tripla, quaecunque autem statuatur refractionis ratio, ipsa eadem hic obtinebit, quicunque scilicet sit valor m & n, & utravis sit earum major, erit $YZ:MN::n:m \infty n$, & hoc primum probe notandum est.

IX.

Sed hoc punctum Y neutiquam verum focum exhibet, propterèa quod recta CR non est in verâ sinus
recti positione; ea est etenim CD: ut cum radio ab
N defluente rectum conficiat anglum ad D; agatur
igitur per D recta NDX, qui verus erit refractus radius,
X verus focus, & ZX vera aberratio.

X.

Dispiciendum est porro unde constet haec YX, quae
in horum schematum priore est auferenda, in altero
vero addenda ipsi ZY: nempe in utroque per parallelas BZ & GRD bina dantur Triangula similia NRI
& NYX, quorum latera NY, NR, seu NX, NI, sunt
in priore schemate, secundum Hugenium, in ratione
triplâ, in posteriore vero, tantum in duplâ, unde in
eadem omnino sunt YX & RI; & si RI dicatur $= e$,
tota aberratio ex horum utroque confecta erit
$nv \pm me/m \backsim n$: hujus autem omnes termini noti
sunt praeterquam ipse e, qui conferendo hanc
aequationem cum prius inventâ in §VII. scilicet
$n^2(v + mny/m^2n^2) = nv \pm me/m \backsim e$, facile erui poterit: nempe adhibendo numeros more Hugeniano
$(m:n::3:2)$

$$e = \begin{Bmatrix} 2/5 & \overline{v-y} \\ 3/5 & \overline{y-v} \end{Bmatrix} ubi \begin{Bmatrix} m \\ n \end{Bmatrix}$$

major. Qui valor tamen, propter laborem eruendi valorem ipsius y, hoc pacto forte facilius investigabitur.

XI.

In utroque schemate propter RG & DN ductas
à G & N, tangentes circulum HDR in punctis R & D,
& per haec puncta ductas rectas CL & CP, est arcus
$LP =$ arcui NG: hic autem arcus est differentia
arcuum, quorum sunt sinus NV & GF, sinus scilicet
incidentiae & refractionis; & semis ejus est LK, cujus
angulus ad centrum est LCK seu RCI; & hujus anguli
ad Radium CR est Tangens $RI = e$. Dicatur igitur,
ut Radius: ad Tangentem anguli LCK $(=1/2$ arc:
$GN)$: ita recta CR $[=$ sinui GF seu $(n/m)NV]$ ad RI
quaesitam: quae cum cognita sit, CR acquiretur
unicâ tantum inspectione Tabularum, ut obtineatur
tangens anguli LCK, seu semissis arcus NG.

XII.

Sed antè dictum est, Hugenium methodum mon-
strasse, aberrationem per surdos eruendi, numerosque
propterea quosdam dedisse absolutos; mihi vero eos
cum methodo suâ perpendenti, quaedam occurrisse
cum lectore communicanda; quod hic in transitu
praestandum censeo, sunt autem ea hujusmodi. Et
pro figurâ suâ pag. 89., usurpabimus praecendentem
12mam; ea enim totam illam Hugenii repraesentat: in

quâ NBV est Lens plano-convexa, cujus centrum C,
axis CBZ Radium lucis axi parallelum excipiens in
planâ superficie NV ad N; ideòque minimè eam refrin-
gens, quem tamen per convexam NB inflectit ad X:
& cum per §. VII. partis primae, hujus lentis focus
primarius sit Z, ideòque aberratio ZX: pro eâ investigandâ, ponit $CB = a$, $NV = b$, & $CX = x$:
quum autem ex constanti refractionis lege, sint CX
& NX in ratione refractionis, h.e. secundum ipsum ut
$3:2$, erit $NX\,(2/3)x$, & ejus quadratum $(4/9)xx$, ex
quâ demptâ bb, erit $VX = \sqrt{(4/9)xx - bb}$: & ex
operatione omninò consimili erit $CV = \sqrt{aa - bb}$, &
tota $CX = \sqrt{(4/9)xx - bb} + \sqrt{aa - bb}$, unde eruit
$x = 3/5\sqrt{9aa - 9bb} + 3/5\sqrt{4aa - 9bb}$: quae resolu-
tio cum sit subobscurior, hic facem imperitioribus
praeferre libet, ut sequitur.

Inventâ $x = \sqrt{(4/9)xx - bb} + \sqrt{aa - bb}$.
Transferendo fiet $x - \sqrt{aa - bb} = \sqrt{(4/9)xx - bb}$.
Quadrando, $(4/9)xx - bb$, $= xx - 2\sqrt{aa - bb} \times$
  $x + aa - bb$.
Auferendo $(4/9)xx - bb$, ex utrâque parte, restat
  $0 = (5/9)xx - 2\sqrt{aa - bb}\,x + aa$.
Hinc aequatio, $(5/9)xx - 2\sqrt{aa - bb}\,x = -aa$.
Dividendo per $5/9$, $xx - 18/5\sqrt{aa - bb}\,x = (9/5)aa$.
Complendo, quadrat. $xx - 18/5\sqrt{aa - bb} \times$
  $x + (81/25)aa - (81/25)bb = (81/25)aa$
  $- (81/25)bb = (9/5)aa$.
Sed est $(9/5)aa = (45/25)aa$.
Itaque erit $xx - 18/5\sqrt{aa - bb}\,x + (81/25)aa$
  $- (81/25)bb = (36/25)aa - (81/25)bb$.
Extrahendo radicem & transferendo,
  $x = \sqrt{(36/25)aa - (81/25)bb} + 9/5\sqrt{aa - bb}$.
Dividendo per $3/5$ seu $\sqrt{9/25}$, $x = 3/5\sqrt{4aa - 9bb}$
  $+ 3/5\sqrt{9aa - 9bb}$.

XIII.

Atque ita se habet Hugenii processus in figurâ suâ
pag. 89, pro radiis axi parallelis, per superficiem
planam ex aëre vitrum ingredientibus, & ex hoc in
illum denuò exeuntibus: sed pro vitrum ingredientibus
per superficiem convexam, sit in figurâ praecedente
11ma. BN superficies convexa, in quam impingit ad
N radius lucis axi parallelus AN, qui per hanc superficiem refringitur ad X, & sint in hâc omnia, ut prius
in alterâ, scil. $CB = a$, $NV = b$, & $CX = x$, & ex
pari ratiocinio, hic erit $NX = (3/2)x$, ejus quadratum
$(9/4)xx$, & ipsa $x = \sqrt{(9/4)xx - bb} + \sqrt{aa - bb}$
unde processu prioris omnino consimili, eruetur
$x = 2/5\sqrt{9aa - 4bb} + 2/5\sqrt{4aa - 4bb}$.

XIV.

Est quidem satis ingeniosa haec Hugenii deductio,
sed aberrationem elicere hâc methodo esset omnino
operosum. Quod autem hic mihi occurrit observan-

dum est, Hugenio, cum haec scriberet, serierum methodum vix notam fuisse, alioqui proclive illi fuisset in hâc suâ aequatione, loco bb, quod est sinus recti quadratum, ex vulgari aequatione circuli, valorem ejus substituisse, nempe (ponendo sinum versum $= v$) $2\ av - vv$, qui valorem ipsius x promptissimè exhibuisset. In aequatione enim suâ, $x = 3/5\sqrt{4aa - 9bb} + 3/5\sqrt{9aa - 9bb}$, quae mutatione quam, dixi factâ, erit $x = 3/5\sqrt{4aa - 18av + 9vv} + 3/5\sqrt{9aa - 18av + 9vv}$: hujus partis radix est $(3/5)3a - 3v$; illius vero non nisi per seriem infinitam est extrahenda, quae erit; $2a - 9v/2 - 45v^2/16a - 405v^3/64a^2 - 16605y^4/1024a^3 - 185895v^5/4096a^4$ &c. cui si addatur prior $3a - 3v$, fiet $5a - 15v/2 - 45v^2/16a - 405v^3/64a^2 - 16605v^4/1024a^3$ &c. cujus $3/5$. partes erunt $3a - 9v/2 - 27v^2/16a - 243v^3/64a^2 - 9963v^4/1024a^3 - 111537y^5/4096a^4$ &c.

XV.

Ita etiam pro fig. 11mâ habuimus

$$x = 2/5\sqrt{9aa - 4bb} + 2/5\sqrt{4aa - 4bb}:$$

in quâ aequatione si pariter pro bb substituatur $2\ av - vv$, erit

$$x = 2/5\sqrt{9aa - 8av + 4vv} + 2/5\sqrt{4aa - 8av + 4vv},$$

cujus postremae partis radix est ipsa $2a - 2v$; & prioris per seriem extrahenda, erit

$$3a - 4v/3 + 10v^2/27a + 40v^3/243a^2 + 110v^4/2187a^3 + 40v^5/19683a^4 \text{ \&c}$$

cui si addatur partis prioris radix $2a - 2v$, fiet $5a - 10v/3 + 10v^2/27a + 40v^3/243a^2$ &c. ut prius; cujus $2/5$ sunt, $2a - 4/3v + 4v^2/27a + 16v^3/243a^2 + 44v^4/2187a^3 + 16v^5/19683a^4$ &c. & hae duae series sunt omnino eaedem, quas exhibet series Newtoniana, si cum Hugenio in primâ ponatur $R = 3$ & $I = 2$, & in posteriori $I = 3$ & $R = 2$, & v pro x: iidem inquam provenient numeri, ut experienti constabit.

XVI.

Nunc etiam de serie illâ Newtoni videamus, in quâ eruendâ (Propos XXXI. page 132) cum processus ejus sit intricatior & minus in Analysi exercitatis aliquid negotii facessere poterit, eam explanandam duxi. Sit in eâdem fig. 12mâ. BN sphaerica superficies, cujus Centrum C, axis CZ, radius axi parallelus AN, & ejus refractus NX, focus sphaerae primarius Z; ideoque hujus refracti aberratio ZX: ad cujus quantitatem determinandam, primò demittantur ad XN productam, itidemque ad CB, normales CD & NV, & dicatur $CB = a$, $VB = v$, & $CX = x$ (habet ille pro $VB\ x$, & pro $CX\ z$) & ex indole circuli erit $NV^q = 2av - vv$, cui addatur $VX^q(= \overline{x - a + v}^2)$ hoc est, $aa + vv + xx - 2av + 2ax + 2vx$, & prodibit $aa + xx - 2ax$

$+ 2vx$. Cum autem NV & CD sint ut sinus incidentiae & refractionis, seu ut I ad R & propter Triangula similia CDX & NVX, NX & CX sint in eadem ratione: erit $I^2:R^2::(NX^2: CX^2)\ \overline{aa + xx - 2ax + 2vx}: xx$; adeoque $I^2xx^2 = R^2 \times \overline{aa + xx - 2ax + 2vx}$; & factâ reductione, inquit Autor,

$$xx = \frac{2R^2ax - 2R^2vx - R^2aa}{R^2 - I^2},$$

extractâque radice

$$x = \frac{R^2a - R^2x + R\sqrt{I^2aa - 2R^2av + R^2vv}}{R^2 - I^2}$$

quod hoc modo fiet.

XVII.

Positis $m = I$, & $n = R$, penultima Autoris aequatio ita se habebit $m^2x^2 = n^2aa + n^2xx - 2n^2ax + 2n^2vx$: sed hic, cum major sit n quam m, ut fiat deductio, nempe $n - m$, omnia signa sunt mutanda, unde deveniet ex partium translatione $n^2x^2 - m^2x^2 = 2n^2ax - 2n^2vx - n^2aa$, & hinc

$$xx - \frac{2n^2ax + 2n^2vx}{n^2 - m^2} = -\frac{n^2aa}{n^2 - m^2}:$$

Complendo quadratum, extrahendo Rad. & Transfdo, erit

$$x = \sqrt{\frac{-n^2aa}{n^2 - m^2} + \frac{n^4a^2 - 2n^4av + n^4v^2}{n^4 - 2n^2m^2 + m^4}} + \frac{n^2a - n^2v}{n^2 - m^2}[.]$$

Ducendo $n^2 - m^2$ in numeratorem $-n^2aa$ ut fiat communis Denominator

$$x = \sqrt{\frac{-n^4aa + n^2m^2aa + n^4aa - 2n^4av + n^4vv}{n^4 - 2n^2m^2 + m^4}}$$
$$+ \frac{n^2a - n^2v}{n^2 - m^2}:$$

inde elisis, quae se invicem destruunt, diviso toto numeratore per n^2, & praefixâ radice suâ n signo radicale, extractâ Denominatoris radice, & inde toto numeratore ad eundem Denominatorem reducto, datur Autoris

$$z = \frac{n^2a - n^2v + n\sqrt{m^2aa - 2n^2av + n^2v^2}}{n^2 - m^2}:$$

Extrahenda porro est radix partis, quae est sub signo radicali, per seriem infinitam, eritque ea hujusmodi

$$ma - \frac{n^2v}{m} \begin{Bmatrix} -\dfrac{n^4}{2m^3} \\ \\ +\dfrac{n^2}{2m} \end{Bmatrix} \frac{v^2}{a} \begin{Bmatrix} -\dfrac{n^6}{2m^5} \\ \\ +\dfrac{n^4}{2m^3} \end{Bmatrix} \frac{v^3}{a^2} \begin{Bmatrix} -\dfrac{5n^2}{8m^7} \\ +\dfrac{6n^6}{8m^5} \\ -\dfrac{n^4}{8m^3} \end{Bmatrix} \frac{v^4}{a^3}$$

$$\left. \begin{array}{l} -\dfrac{7n^{10}}{8m^9} \\[2mm] +\dfrac{10n^8}{8m^7} \\[2mm] -\dfrac{3n^6}{8m^5} \end{array} \right\} \dfrac{v^5}{a^4} \qquad \left. \begin{array}{l} -\dfrac{21n^{12}}{16m^{11}} \\[2mm] +\dfrac{35n^{10}}{16m^9} \\[2mm] -\dfrac{15n^8}{16m^7} \\[2mm] +\dfrac{n^6}{16m^5} \end{array} \right\} \dfrac{v^6}{a^5}$$

In hujus seriei unumquemque numeratorem injiciatur illa n, quae praefigitur signo radicali; pertinet enim ad totam seriem; addantur prioribus etiam duobus membris, ipsa $+ n^2 a$, & $- n^2 v$, quae radicali connectuntur; & dividatur quodque membrum per Denom. communem $n^2 - {}^2m$ $[n^2 - m^2?]$, itaque primo habebitur $n^2 a + nma / n^2 - m^2 = na / n - m$; pro secundo

$$\frac{(-n^3 v / m - n^2 v)}{n^2 - m^2} = \frac{-n^3 v - n^2 m v}{n^2 m - m^3} = \frac{-n^2 v}{nm - m^2};$$

pro tertio $[(-n^5 / 2m^3) - (n^3 / 2m)] / n^2 - m^2$; quod ut ritè ordinetur, ducatur m^2 in 2dae partis numeratorem, ut utriusque partis communis Denominator fiat $2\,m^3$; & ducatur hic in communem Denominatorem $n^2 - m^2$, unde totum hoc membrum deveniet $-n^5 + n^3 m^2 v^2 / 2m^3 n^2 - 2m^5 a$; quod divisum more communi dabit quotientem $n^3 v^2 / 2m^3 a$. Atque eodem modo per totam seriem pergendo tandem habebitur haec.

$$\frac{na}{n-m} - \frac{n^2 v}{mn - m^2} - \frac{n^3 v^2}{2m^3 a} - \frac{n^5 v^3}{2m^5 a^2} - \frac{5n^7 - n^5 m^2 v^4}{8m^7 a^3}$$
$$- \frac{7n^9 - 3n^7 m^2 v^5}{8m^9 a^4} - \frac{21n^{11} - 14n^9 m^2 + n^7 m^4 v^6}{16m^{11} a^5}.$$

XVIII.

Quam seriem eò longius hic deduxi, non quod illius adeò provectae aliquis futurus sit usus, sed quò clarius ex ipso intuitu videantur ea, quae de illâ jam dicenda sunt. Dixi enim in §III. praeced., eam, uti ab illustri Autore traditur, duplice laborare defectu, quorum hic est unus; quod cum ex ejus &c. quod subjicitur quarto membro, in quo apud eum desinit illa series, jure quivis expectaret, eodem ordine canonico processuram eam, scilicet, quod foret proximum membrum $n^7 v^4 / 2m^7 a^3$; quod tamen longè secus est, & sic in sequentibus: Alter horum est, quod notasse oportuit Autorem, haec signa quae ab illo, atque etiam suprà sunt tradita, inservire tantum, cum angulus radii refracti major est angulo incidentis, unde R seu n major est quam I aut m: hoc est, cum radii Lucis e medio densiore in rarius exeunt: quum autem contra fit, omnia signa post secundum mutanda sunt, unde pro his radiis, series erit

$$\frac{ma}{m-n} - \frac{n^2 v}{m^2 - mn} + \frac{n^3 v^2}{2m^3 a} + \frac{n^5 v^3}{2m^5 a^2} \ \&c.$$

& dempto membro priore, restabit

$$\text{aberratio} = \frac{n^2 v}{m^2 - mn} - \frac{n^3 v^2}{2m^3 a} - \frac{n^5 v^3}{2m^5 a^2} \ \&c.$$

Constat enim ex figuris 13iâ. & 14tâ., aberrationem in hoc casu, scilicet fig. 13iâ. semper minui in altera verô augeri. Et si pro speciebus usurpentur numeri, more Hugeniano, scilicet pro $I:R$, seu $m:n$, $2:3$, vel $3:2$, prout exigit rei ratio; iidem planè in utroque casu exurgent numeri, & omnino his gemini, quos habuimus ex figurâ Hugenianâ deductos §. §. XIV. XV.

XIX.

Prius etiam quam hoc argumentum de radiis axi parallelis dimittatur, cùm Hugenii regulas susceperim demonstrandas, aequum est, ut de illis, quos tradidit, numeris, majori compendio in errorum investigatione inservientibus, & de illorum inventione, etiam aliquid dicatur: quorum priores, qui occurrunt, sunt (pag. 90.) 9/2 crassitudinis, seu sinus versi convexitatis vitri, pro errore radiorum e densiore medio in rarius exeuntium, & (pag. 92.) 4/3 ejusdem, pro densius ex aëre ingredientibus; etiamque (pag. 91.) pro totâ Lente plano-convexâ, cujus facies convexa radiis lucis obvertitur, 7/6 ejusdem crassitudinis, itidem etiam pro lente pariter convexa (pag. 94.) 5/3 crassitudinis, seu 5/3 sinus versi convexitatis: quae omnia ut vir ille eximius mirandae erat sagacitatis, haud aliter mihi eliciuisse videtur, quam aliquoties rem eandem per Trigonometriam, aut aliter experiundo, quod mihi quidem tentanti in eorum aliquibus obtigit, priusquam investigationem accuratiorem per calculum aggrederer, aut hasce regulas inspexerim. Minimè vero omnino accuratos esse hos numeros, ex praecedentibus abunde patet, in quibus constat, 9/2 & 4/3 primos tantummodo esse numeros, qui è serie Newtonianâ pro diverso valore I & R, seu m & n, exurgunt, itidemque in caeteris.

XX.

At quânam arte vel quali indagine illam eruerit regulam, quae (pag. 94) pro aberratione Lentis imparium convexitatum investigandâ exhibetur; ubi positâ convexitatis radiis obversae semidiametro $= a$, alterius semidiametro $= n$, crassitudine lentis $= q$, errorem statuit $= (27aa + 6an + 7nn/6 \times a + n^2)q$ (ita enim scribi oportuit,) haud perinde facile explicatu censeri possit: non tamen ita in obscuro positum est, quin assequi fortasse valeamus. Quo pacto in primos illos numeros 27 in numeratore $= m^3$, & 6 in Denominatore $= mn$, inciderit, mihi satis compertum est: & cùm in omni ejusmodi tentamine, de extremis, ut norunt periti, semper faciendum sit periculum: horum autem extremorum alterum est Lens utrinque pariter convexa, cujus aberrationem constabat esse 5/3: alterum vero Lens plano-convexa, cujus est error 7/6. Sed quum fieri possit, ut cum altera semidiameter sit unitas, seu = 1, altera etiam ad decies millies tantam vel suprà exurgat, ita temperandi erant numeri, ut si utraque a & n foret unitas, quarum summa = 2, quadratum est 4, & 6 × 4 = 24 (quare vero senarius hic sit usurpandus, mox patebit) ideo pro aequalibus semidiametris numeratorem 40 (scilicet ut 5/3 = 40/24 efficiat) esse oportet. Propter alterum vero extremum (Lentem plano-convexam scilicet, cujus error est 7/6) necesse erat ut n^2 septenarium haberet adjunctum, etiamque ut senarius esset in Denominatore; quod ab exemplo clarius elucescet. Sint igitur a & n aequales, & utravis = 1, & ob causam jam redditam, nempe quod Denom. $6 \times \overline{a + n^2} = 24$, numeri in Numeratore utcunque affixi, debent simul adjuncti conficere 40, ut sint = 5/3, & cum singuli horum aa, an, nn, sint = 1, perinde esset, si poneretur $38aa + an + mn$, aut $aa + 10ann + 29nn$, aut alio quovis modo; dummodo numerorum coefficientium summa quadragenarium non superaret. At si ponatur $a = 1$, n vero = 1000, erit $27aa = 27$, $6an = 6000$, $7nn = 7000000$, quorum summa = 7006027 & $6 \times \overline{a + n^2} = 6012006$, qui duo rite compositi haud longe absunt à 7/6.

XXI.

Ex eâdem autem aequatione (pag. 95) aliam eruit pro investigando errore ex datis semidiametro a superficiei exterioris, & Lentis foco, quem ponit = d, unde cum per §XVI. in priore hujus, sit $d = 2an/a + n$, erit per legitimam reductionem $n = ad/2a - d$, quo ubique ritè substituto in locum n in regulâ priori $(27aa + 6an + 7nn/6 \times \overline{a + n^2})q$ proveniet aberratio $= (27aa - 24ad + 7dd/6aa)q$; quae licet à priore diversa videatur, cum tamen inde deducatur, & eidem innitatur fundamento, eadem omnino censenda est.

XXII.

His ita absolutis, jam de radiorum lucis obliquè in superficies sphaericas incidentium aberrationibus videndum erit, quas nemo, ut prius dictum est, quantum novi, adhuc in calculum vocavit, nedum pertractavit. In quibus investigandis ad hunc modum pergere licebit.

Sit in fig. 15. superficies sphaerica convexa BN, cujus semidiameter $CB = CN$, axis AX; sit AN radius Lucis obliquè in eam impingens ad N, arcûs BN sinus rectus NV, sinus versus BV. Per §. X. partis prioris, si producatur AN ad D, eique a Centro C agatur normalis CD, eâque divisâ in ratione refractionis, per punctum divisionis agatur arculus dEf, & ad hunc ducatur tangens NEX, occurrens axi in puncto X, hoc punctum X erit radii obliqui ad N incidentis, & inde refracti verus focus, cujus aberrationem XZ a puncto Z, ejusdem superficiei foco primario per §. X. inveniendo, per aliquam ex methodis sequentibus investigare licebit.

XXIII.

Dixit, ut ante notatum est, Clariss. Newtonus, "Eodem modo, quo errores parallelè incidentium determinantur, consimiles divergentium, vel convergentium (h.e. obliquorum) licet calculo difficiliori determinari posse." Hujus autem calculi difficultas tota, tum in hoc constat, quod loco radii paralleli, cujus longitudinis, utpote infinitae, nulla habetur ratio, hic habenda est non tantum puncti A, unde proveniunt, vel ad quod convergunt radii, distantia a vertice B; sed si accuratius instandum fuerit, etiam ipsa longitudo radii AN: tum etiam (quod majoris momenti est;) in hoc, quod ratio refractionis minimè à sinu recto NV sumenda sit, sed a rectâ CD; quod §. XXIII. & eas, quae sequuntur in priore parte, rite perpendenti abundè patebit.

XXIV.

Cum vero in his quae jam traduntur, variae exhibeantur methodi radiorum axi parallelorum, sed remotius ab eo incidentium errores determinandi, idem hic per quamvis illarum fiet; Radium lucis AN tantummodo in calculum inducendo, ad hunc modum; unde varietas, quae exinde orietur, ultrà perpendi poterit.

XXV.

Ponatur ideò Radii AN longitudo fig. 15. $(= \sqrt{AV^2 + VN^2}) = a$, $CB = r$, $AC = c$, $NV = d$, sinus versus BV (recti NV) = v, sinus versus EG (recti CE) = y, unde erit $CV = r - v$, $NE = r - y$, & dicatur $CX = x$. Cum autem similia sint Triangula AVN & ADC, erit $AN:NV::AC:CD$. h.e. $a:d::c:dc/a$, unde erit $CE = dnc/ma$: etiamque $NV:VX::CE:EX$. h.e. $d: x + r - v:: ndc/ma: nc/ma \times \overline{x + r - v}$: hinc $nc/ma \times \overline{x + r - v + r - y} = NX$. Sed etiam $CE:CX::VN:NX$; h.e. $ndc/ma: x::d: max/nc$: hinc aequatio $nc/ma \times \overline{x + r - v} + r - y = max/mc$; in quâ si pro ma/c substituatur s. erit $sx/n = n/s \times \overline{x + r - v + r - y}$, ex quâ rite ordinatâ, ut in §. VII. proveniet $x = nr/s - n - n^2v + sny/s^2 - n^2$, cujus pars postrema vera est aberratio, & eodem modo ut in eâ sectione dictum est, resolvi poterit in has partes $(n/2)(v + y/s - n) \pm nv \backsim y/2s + n$. Porrò de primâ parte hoc animadvertendum occurrit, nempe si ponatur a = distantiae puncti A â vertice Lentis B, cum sit $s = ma/c$, haec $nr/s - n$ veram exprimet distantiam foci primarii per §. V. inveniendi a Centro C.

XXVI.

Sed methodo, aliquantò (ut quibusdam fortasse videbitur) faciliore, sine substitutione, & solas litteras m, & n, usurpando, confici poterit hoc problema: scilicet in fig. 16tâ. si recta $CE = ndc/ma$ in §. & figurâ praecedente, agatur non a centro C, sed a puncto N, & eâdem distantiâ quâ prius, ducto arculo dEf; huic a centro C agatur tangens CES: namque in Triangulo inde confecto CSD, omnia se habebunt, ut in Triangulis NXV in fig. 11ma. & 12mâ. ad §. IV. namque hic erit SN ad SC in ratione refractionis, ut ibi fuit XC ad XN, seu $n:m$ unde positâ $CN = r$, & $NS = x$, per calculum omnino illius consimilem, qui exhibetur in §. §. V. VI. habebitur ipsa x; posito scilicet sinus versos in duplicatâ ratione esse ad suos rectos, nempe $x = (n/m - n)r - (n^2/m^2 - mn)v$. aut si majore accuratione agendum & ipsa y sit usurpanda, erit, ut in §. VII. $x = (n/m - n)r - n^2v + mny/m^2 - n^2$, cujus pars postrema resolvi poterit, ut jam saepius dictum est. Inventâ hoc pacto quantitate hujus $x = NS$, & ductâ CS, agatur etiam huic parallela NX, unde emergent bina Triangula omnino similia ASC & ANX: detractâ autem SN ex AN, erit $AS:SN::AC:CX$, unde habebitur idem punctum X ut prius. Nihil enim interest, an eadem distantia erigatur à C, an demittatur ab N; namque ut rectae utrinque ductae sint parallelae omnino necesse est.

XXVII.

Sed si cui magis arriserit methodus per seriem, inveniet in Clariss. Newtoni Arithmeticâ Universali Problema, pro radiorum divergentium focis inveniendis, quod in primâ praeclari illius operis editione, Anno 1707. est Prob. 17. sed in 2dâ. Prob. 31. unde elici ad hanc rem poterit series, licet non citra laborem oppidò nimium, si minore quovis modo confici possit; quod quidem ex praecendentibus rite animadversis haud difficulter, ni fallor, expedietur. Dicatur enim in praecedente fig. 15tâ. ut prius, $AN = a$, $AC = c$, $CN = CB = r$, $BV = v$, $CX = x$, sed nunc sit $NV = \sqrt{2rv - vv}$, & propter similia Triangula ANV, ACD, erit $CD = c/a\sqrt{2rv - vv}$, ideoque $CE = nc/ma\sqrt{2rv - vv}$. Erit etiamque ut prius $VX = x + r - v$, & hujus quadratum $= xx + rr + vv + 2rx - 2vx - 2rv$, cui addatur quadr. $NV = 2rv - vv$, unde habebitur $\sqrt{xx + rr - 2rx - 2vx} = NX$, Item propter similia Triangula NXV, CXE; erit

$$NX:NV::CX:CE = \frac{x\sqrt{2rv - vv}}{\sqrt{xx + rr + 2rx - 2vx}}.$$

Sed prius inventa erat $CE = nc\sqrt{2rv - vv}/ma$, unde orietur aequatio, ex cujus numeratoribus eliminatâ utrinque communi radicali $\sqrt{2rv - vv}$, restabit $x/\sqrt{xx + rr + 2rx - 2vx} = nc/ma$ & per crucem facta multiplicatione $max = nc\sqrt{xx + rr + 2rx - 2vx}$, seu $(ma/nc)x = \sqrt{xx + rr + 2rx - 2vx}$, dicatur ut prius $ma/c = s$, & totum quadrando erit $(ss/n^2)xx = xx + rr + 2rx - 2vx$. & rite ordinando ut in §. §. XLVI. & XLVII. in casu omnino consimili monstratum est, dabitur $x = n^2r - n^2v + n\sqrt{s^2n^2 - 2n^2rv + n^2v^2}/s^2 - n^2$ unde eodem modo, quo in §XVII. proceditur, eruetur

haec series isti omnino similis, praeterquam quod hic vice m usurpatur $s = ma/c$, & quod hic de radiis medium densius subeuntibus agitur, scilicet

$$x = \frac{s}{s - n}r - \frac{n^2v}{s^2 - ns} - \frac{n^3v^2}{2s^3r}$$
$$- \frac{n^5v^3}{2s^5r^2} - \frac{5n^7 - n^5s^2}{8s^7r^3}v^4 \text{ \&c.}$$

Notandum porro, in his omnibus dari methodos non tantum errores refractionum investigandi, sed absolutos ipsos focos radiorum quorumvis remotius ab axe impingentium qui foci si conferantur cum illis, quorum traditur investigatio in priore hujus parte, inde facile aberrationis quantitas eruetur.

XXVIII.

Porro eodem planè modo, quo obliqui radii AN focus investigatur, etiam ipsius NX (ratione habitâ diversorum mediorum) focus est investigandus, unde habebitur focus totius Lentis; quo collato cum foco per regulas Hugenianas parte priore demonstratas invento, exinde facillimè deprehendetur cujusvis radii in quavis Lente totus error.

CONCERNING THE ABERRATIONS FROM THE PRIMARY FOCUS OF THE LIGHT RAYS WHICH FALL ON SPHERICAL SURFACES FARTHER FROM THE AXIS. Translated by Benedict Monostori.

I.

Since this very useful and entertaining science, which is named Dioptrics, is one composed of a mixture of physics and mathematics; therefore, that part of it which leans on physical reasonings hardly allows the rigor of geometry in some [aspects]. To be sure, the illustrious gentleman, Chris. Huygens, expounded it in his excellent work in such a way that to every user at all it could look satisfactory, even if the rules which he gave considerably lacked that same rigor. Whoever, therefore, wishes to put some further effort in such studies would seem to be wasting his time. But when, after discovering what in his rules was to be given by the clear light of geometry, and also having more deeply pondered over those [things] which, in addition, the author wrote concerning the aberrations [caused] by lenses on the rays [incident] farther from the axis, and having examined more accurately from where and by what calculation he arrived at his absolute numerical values which he recommends to use, certain things have occurred to me which illustrate the very matter not a little and the communication of which I think the zealous reader will hardly mind.

II

Where these aberrations—as I understand them here—come from, has been abundantly demonstrated in §§ IV and V of this first part; namely, since in fig. 2 the real or proper focus of the surface BML is point Z, that of the ray AM, which is farther from the axis, is S, and of ray OL is T, whose distances from point Z, i. e., ZS, ZT, are the very aberrations which are discussed here.

III.

As I know, no one among the writers about Dioptrics mentioned these investigations except this one [Huygens], nor did they say anything except that in illustrious Newton's Optical Lectures, which he delivered at one time at Cambridge and which were published recently, there occurs a proposition (Propos. XXXI, Page 132) to determine the focus of parallel rays which fall farther from the axis on a sphere and are refracted hence. He solved this by a rather complicated method through an infinite series evolved from surds. This series, however, although it is good enough as the author uses it there, labors under two defects as will be abundantly clear from the following. Neither of these [authors], however, says anything about investigations of errors due to obliques, either divergent or convergent, though the knowledge of these is of even greater importance in Dioptrics than that of the right angle or the parallel ones. For in all composite optical instruments the rays fall obliquely on more surfaces of the lenses than at right angles: in a telescope of two lenses, if they fall parallel on the first [surface] of the so-called objective, they hit obliquely the other three: if there are four lenses, they can reach two parallel, but they will reach all the other six [surfaces] obliquely: and in the microscopes, always obliquely. Newton himself says in the Scholion, to the mentioned proposition on page 136, that in the same way as he determined the errors due to parallel incidence, similar ones due to divergent or convergent [rays] could be determined but through a more difficult calculation, of which more later.

IV.

But Huygens himself shows in Proposition XXVII, a method through surds, and a complicated and difficult one, for the investigation of these errors. But to spare the tedious labor, he gives numerical values for the parallel [rays] incident on all kind of lenses, about which we shall see in the following. For the first among them, namely, suppose that the versed sine $= v$ for the rays passing out of the glass into the air, he says the error will be $9/2v$, conversely [for those] from the air into glass $4/3v$: but these numbers are found to be true no further than a duplicate ratio holds between the right and versed sines which hardly exceeds one or two degrees. But the whole issue, if I am not mistaken, will be resolved this way much more clearly and easily.

In figs. 11 and 12, let BN be a spherical surface convex toward M, whose center is C, its axis CB; let AN be a light ray incident at N parallel to the axis of the surface, in fig. 11 from the air into the denser medium, in fig. 12 from the latter into the air. According to the rules given in the first part, if this ray would arrive very near to the vertex B, its focus would fall on the axis at point Z: since it would be $ZB:ZC::I:R$. But now, as it was demonstrated in §VI of the first part, it must fail to reach that focus. Assume, therefore, on [the line of] right sine NV that $VN:VR::I:R$, with the interval $CD = VR$, construct from the center C an arc eDf, to which draw a tangent NDX from N, and the true focus of this ray NX will be at the point X.

V

To determine this focus by calculation, put the radius $BC (= NC) = a$, the sine $VN = m$, $CD (= VR) = n$: and since $VN:VR::I:R$, also $I:R::m:n$. Let the versed sine $BV = v$, then it follows that $CV = a - v$; and since the versed sines, in such small arcs, are in a duplicate ratio of their right sines, the versed sine of the right one CD will be $(n^2/m^2)v$; hence ND also will be $= a - (n^2/m^2)v$: let further $CX = x$. The triangles NVX, CDX are similar because of the right angles at V and D, and the common one at X, hence $VN:VX::CD:DX$; i.e. in fig. 11 $m:x + a - v::n: nx + na - nv/m = DX$; and from this

$$n\frac{\overline{x + a - v}}{m} + a - \frac{n^2}{m^2}v = NX,$$

but $NV:CD::NX:CX$, i.e.

$$m:n::\frac{n}{m}\overline{x + a - v} + a - \frac{n^2}{m^2}v:x,$$

hence the equation $mx = (n^2/m^2)\overline{x + a - v} + na - (n^3/m^2)v$, which properly rearranged will give $x = (mn + n^2/m^2 - n^2)a - [n^2 + (n^3/m)/m^2 - n^2]v$. Divide out its first part by $m + n$; since $m^2 - n^2 = \overline{m + n} \times \overline{m - n}$; and bring m into each part of the second, and divide by $m + n$; then it will be $x = n/(m - n)a - (n^2/m^2 - mn)v$.

VI.

In fig. 12, VX will be $= x - a + v$ (if supposed, as before, $CX = x$) and $NX = (n/m)\overline{x - a + v} - a + (n^2/m^2)v$, and if we proceed in the same manner as above, and the signs are changed throughout, the very same equation will come out (supposed that here n be the greater term of the ratio), that is $x = (n/n - m)a - (n^2/mn - m^2)v$. Taking in each of these equations, according to Huygens, $m:n = 3:2$ or $2:3$, we shall have exactly the same numerical values as he, namely, in this latter one $3a - (9/2)v$, and in the first one $2a - (4/3)v$. They give even the very terms of the Newtonian series which will shortly be discussed. For the first term $(n/m - n)a$ is CZ itself:

the second one $(n^2/mn - m^2)v$ is the aberration itself, namely ZX, to be taken out of CZ. It will resolve, to be sure, into these factors $n/m - n \times n/m$, which according to the value of m and n will be either $2/1 \times 2/3 = 4/3$ or $3/1 \times 3/2 = 9/2$.

VII

But in the previous process it was taken for granted that the versed sines DT, in both figures, were in a duplicate ratio of their right sines VN and VR or CD; which, however, beyond the first or second degree of the quadrant, hardly holds. Therefore a more universal rule must be looked for: which we shall obtain very easily from the same figures if we substitute for $(n^2/m^2)v$ $y = DT$, whose value will be found in the tables of sines, by deriving namely from the radius ND itself, which is the sine of the complement of CD itself, and thus the equation will be given $mx = (n^2/m)x + a - v + na - ny$, whence from a calculation entirely similar to the preceding one will follow $x = (n/m - n)a - n^2v + mny/m^2 - n^2$, whose last part will be the true value of the aberration for parallel rays through all degrees of the circle in which refraction can occur. It could be factored into $(nv + my) \times n/m^2 - n^2$, or also into $n/2 \times v + y/m - n \pm v \backsim y/m + n$. A minus sign should be used, however, when n is the smaller part of the ratio; plus sign when it is greater.

VIII.

But let us consider for greater clarity of the issue where does this aberration ZX come from. That not only will be crystal clear from this geometrical construction, but through it there will be displayed also a more direct and perhaps somewhat easier method to compute and to determine the same. Let in figs. 13 and 14, therefore, AM be a light ray parallel to the axis coming to the surface BNL which is convex toward M from the rarer medium into the denser in fig. 13, contrariwise in fig. 14. If this ray falls very near to the vertex B, assuming that CR is in the ratio of refraction to the sine $NV = MB$ and drawing MRX through point R on the normal CL, [then] point Z will be given (from what has already been demonstrated in the first part) as the primary focus of this surface: but now the ray hits at N: let be drawn, therefore, from N through the same point R the straight [line] NRY, meeting the axis at Y; whence, because BZ and AN are parallel, follows $ZY:MN::$ $ZR:MR::YR:NR$, i.e., in fig. 13 (assuming Huygens's ratio $I:R::3:2$) YZ will be the double of the versed sine MN; and in fig. 14, its triple, to be sure, whatever ratio of refraction is stated, the very same will hold here, namely whatever be the value of m and n, and whichever of them be greater, it will be $YZ:MN::n:$ $m \backsim n$, and this first must be kept in mind.

But this point Y does not represent the true focus at all, because the straight line CR is not in the true position of a right sine; it is, indeed, CD: to make a right angle at D with the ray coming from N; therefore, the straight line NDX should be drawn through D which will be the true refracted ray, X the true focus, and ZX the true aberration.

X.

It should be considered further where does this YX come from, which in the first of these schemes has to be subtracted from, in the second, however, must be added to ZY itself: namely in both through the parallels BZ and GRD two similar triangles NRI and NYX are given whose sides NY, NR, or NX, NI are in the first scheme, according to Huygens, in triple ratio, in the latter, however, only in double, hence in exactly the same [ratio] are YX and RI; and if RI is said to equal e, the whole aberration will be made up from both of them as $nv \pm me/m \backsim n$: but all the terms of this are known outside of e itself, which could be computed easily by comparing this equation with the one found before in § VII, that is $n^2v + mny/m^2 - n^2 = nv \pm me/m \backsim e$: namely using numerical values in the Huygenian manner $(m:n::3:2)$

$$e = \begin{cases} 2/5 & \overline{v - y} \\ 3/5 & y - v \end{cases} \text{ where } \begin{cases} m \\ n \end{cases}$$

is greater. Which value, however, because of the labor [involved] in finding the value of y itself, will be easier perhaps to investigate in this way.

XI.

In both schemes since RG and DN are drawn from G and N, touching the circle HDR in points R and D, and the straight lines CL and CP are drawn through these points, arc $LP =$ arc NG: this arc, to be sure, is the difference of the arcs, whose sines are NV and GF, that is, the sine of incidence and of refraction; and its half is LK whose angle at the center is LCK or RCI [resp.], and the tangent of this angle to the radius CR is $RI = e$. Let us say, therefore, that as the radius: to the tangent of the angle LCK ($= 1/2$ arc: GN): so the straight line CR [$=$ to sine GF or (n/m) NV] to RI under question: if this is known, CR will be acquired merely by a single inspection of the tables to obtain the tangent of angle LCK, or the half of arc NG.

XII.

But it was said before, that Huygens had shown a method to evaluate the aberration through surds, and therefore had given some absolute numerical values; but when I was investigating them with his method, there occurred to me something worthy of communication to the reader; which I judge should be presented here briefly: for his figure on Page 89,[28] we will use the 12th above; for it represents entirely that of Huygens: in which NBV is a plano-convex lens, whose center C, axis CBZ, receiving the light ray parallel to the axis on the plane surface NV at N; therefore, refracting it none, but through the convex [side] NB bends it to X: and since by §VII of the first part, the primary focus of this lens is Z, therefore the aberration ZX: to investigate it, he puts $CB = a$, $NV = b$, and $CX = x$: but, since from the constant law of refraction CX and NX are in the ratio of refraction, i.e., according to him as $3:2$, NX will be $(2/3)x$, and its square $(4/9)x^2$, from which b^2 taken, vx will be $= \sqrt{(4/9)x^2 - b^2}$: and by an entirely similar operation CV will be $= \sqrt{a^2 - b^2}$: and the whole $cx = \sqrt{(4/9)x^2 - b^2} + \sqrt{a^2 - b^2}$ whence he derives $x = 3/5\sqrt{9a^2 - 9b^2} + 3/5\sqrt{4a^2 - 9b^2}$: which solution being somewhat more obscure, let me here present some light to the non-experts, as follows:

Having found that $x = \sqrt{(4/9)x^2 - b^2} + \sqrt{a^2 - b^2}$.

Transferring becomes $x - \sqrt{a^2 - b^2} = \sqrt{(4/9)x^2 - b^2}$.

Squaring, $(4/9)x^2 - b^2, = x^2 - 2\sqrt{a^2 - b^2} \times x + a^2 - b^2$.

Subtracting $(4/9)x^2 - b^2$ from both parts, remains $0 = 5/9x^2 - 2a\sqrt{a^2 - b^2} x + a^2$.

Hence the equation, $(5/9)x^2 - 2\sqrt{a^2 - b^2} x = -a^2$.

Dividing by $5/9$, $x^2 - 18/5\sqrt{a^2 - b^2} x = (9/5)a^2$.

Completing the square $x^2 - 18/5\sqrt{a^2 - b^2} \times x + (81/25)a^2 - (81/25)b^2 = (81/25)a^2 - (81/25)b^2 = (9/5)a^2$.

But $(9/5)a^2 = (45/25)a^2$.

Therefore, it follows $x^2 - 18/5\sqrt{a^2 - b^2} x + (81/25)a^2 - (81/25)b^2 = (36/25)a^2 - (81/25)b^2$.

Taking the root and transferring,
$x = \sqrt{(36/25)a^2 - (81/25)b^2} + 9/5\sqrt{a^2 - b^2}$.

Dividing by $3/5$ or $\sqrt{9/25}$, $x = 3/5\sqrt{4a^2 - 9b^2} + 3/5\sqrt{9a^2 - 9b^2}$.

XIII

Now, such is Huygens's process in his figure on page 89 for the rays parallel to the axis, which enter the glass from the air through a plane surface, and

[28] Logan's references through §XXI of this essay are to "Dioptrica, Pars Secunda, De aberratione radiorum a foco," in *Christiaan Huygens opuscula posthuma* . . . (Leyden, 1703). Huygens's work is more easily found in *Œuvres Complètes de Christiaan Huygens Publiées par la Société Hollandaise des Sciences* 13, 1 (La Haye, 1916), cited hereafter as Huygens, *Œuvres Complètes* 13, 1. The page 89 cited by Logan is in the 1703 edition. It will be found in Huygens, *Œuvres Complètes* 13, 1: p. 285. Logan appears to be referring to Huygens's figure 13.

leave it again for the latter: but for those entering the glass through a convex surface, let BN be in the previous figure 11 a convex surface, on which falls at N a light ray AN parallel to the axis, which is refracted through this surface to X, and let in this [case] everything be as before in the other, namely $CB = a$, $NV = b$, and $CX = x$, and from an identical reasoning, here NX will be $= (3/2)x$, its square $(9/4)x^2$, and x itself $= \sqrt{(9/4)x^2 - b^2} + \sqrt{a^2 - b^2}$ whence by a process altogether similar to the previous will be derived $x = 2/5\sqrt{9a^2 - 4b^2} + 2/5\sqrt{4a^2 - 4b^2}$.

XIV

Though this deduction by Huygens is ingenious enough, still it would be altogether tiresome to extract the aberration by this method. But what occurs to me here should be noted that the method of series was hardly known by Huygens when he wrote this, otherwise it would have been easy for him to substitute in this equation of his instead of b^2 which is the square of right sine, from the commonplace equation of the circle its value, that is (supposing the versed sine $= v$) $2av - v^2$, which would give the value of x itself very promptly. For in his equation $x = 3/5\sqrt{4a^2 - 9b^2} + 3/5\sqrt{9a^2 - 9b^2}$, which after the change I have mentioned being done will be $x = 3/5\sqrt{4a^2 - 18av - 9v^2} + 3/5\sqrt{9a^2 - 18av + 9v^2}$: the root of this [latter] part is $3/5(3a - 3v)$: of the other, however, only through an infinite series can be extracted, which will be:

$$2a - \frac{9v}{2} - \frac{45v^2}{16a} - \frac{405v^3}{64a^2} - \frac{16605y^4}{1024a^3} - \frac{185895v^5}{4096a^4}$$

etc. If the previous $3a - 3v$ is added to it, it becomes

$$5a - \frac{15v}{2} - \frac{45v^2}{16a} - \frac{405v^3}{64a^2} - \frac{16605v^4}{1024a^3}$$

etc. whose $3/5$th part will be $3a - 9v/2 - 27v^2/16a - 243v^3/64a^2 - 9963v^4/1024a^3 - 111537v^5/4096a^4$ etc.

XV

Thus we have also for fig. 11, $x = 2/5\sqrt{9a^2 - 4b^2} + 2/5\sqrt{4a^2 - 4b^2}$: If $2av - v^2$ is substituted for b^2 in this equation as well, it will be

$$x = 2/5\sqrt{9a^2 - 8av + 4v^2} + 2/5\sqrt{4a^2 - 8av + 4v^2}$$

the root of its last part is $2a - 2v$ itself; and [that] of the first extracted by series will be

$$3a - \frac{4v}{3} + \frac{10v^2}{27a} + \frac{40v^3}{243a^2} + \frac{110v^4}{2187a^3} + \frac{40v^5}{19683a^4}$$

etc. If the root of the previous part $2a - 2v$ is added to it, it becomes $5a - 10v/3 + 10v^2/27a + 40v^3/243a^2$ etc. as before; whose $2/5$th is $2a - 4/3v + 4v^2/27a$

$+ 16v^3/243a^2 + 44v^4/2187a^3 + 16v^5/19683a^4$ etc., and these two series are altogether the same, as what the Newtonian Series presents, if with Huygens $R = 3$ and $I = 2$ is put in the first, and $I = 3$ and $R = 2$ in the latter; and v for x: I say, the same numbers will result as it will be clear to the expert.

XVI

Now let us see also about that series of Newton, Since his process in its evaluation is more involved and it might offer some trouble to those less practiced in analysis, I feel I should explain it. Let in the same fig. 12, BN be a spherical surface, whose center [is] C, CZ the axis, AN a ray parallel to the axis, and NX its refracted, Z the primary focus of the sphere: therefore, ZX the aberration of this refracted [ray]: to determine its size, let first the normals CD and NV be constructed drawn to XN and in the same way to CB, and let us call $CB = a$, $VB = v$, $CX = x$ (he has x for VB and z for CX) and from the nature of the circle it follows that $NV^2 = 2av - v^2$, to which should be added $VX^2 = (x - a + v)^2$, i.e., $a^2 + v^2 + x^2 - 2av - 2ax + 2vx$, and it will come out $a^2 + x^2 - 2ax + 2vx$. But since NV and CD are as the sines of incidence and refraction, or as I to R and because of the similar triangles CDX and NVX, NX and CX are in the same ratio: follows $I^2 : R^2 : : (NX^2 : CX^2)a^2 + x^2 - 2ax + 2vx : x^2$; and thus $I^2 \times X^2 - R^2 \times a^2 + x^2 - 2ax + 2vx$; and after the reduction, says the author,

$$x^2 = \frac{2R^2ax - 2R^2vx - R^2a^2}{R^2 - I^2}$$

and its root having been taken

$$x = \frac{R^2a - R^2x - R\sqrt{I^2a^2 - 2R^2av + R^2v^2}}{R^2 - I^2}$$

which is done this way.

XVII

The last but one equation of the author, having put $m = I$, and $n = R$, will become $m^2x^2 = n^2a^2 + n^2x^2 - 2n^2ax + 2n^2vx$: but here, since n is greater than m, to have deduction, i.e., $n - m$, all the signs have to be changed whence comes from the transfer of the parts $n^2x^2 - m^2x^2 = 2n^2ax - 2n^2vx - n^2a^2$, and hence

$$x^2 - \frac{2n^2ax + 2n^2vx}{n^2 - m^2} = \frac{-n^2a^2}{n^2 - m^2}:$$

Completing the square, taking the root and transferring it will be

$$x = \sqrt{\frac{-n^2a^2 + n^2m^2a^2}{n^2 - m^2} + \frac{n^4a^2 - 2n^4av + n^4v^2}{n^4 - 2n^2m^2 + m^4}} + \frac{n^2a - n^2v}{n^2 - m^2}.$$

Introducing $n^2 - m^2$ into the numerator $-n^2a^2$ to have common denominator

$$x = \sqrt{\frac{-n^4a^2 + n^4a^2 - 2n^4av + n^4v^2}{n^4 - 2n^2m^2 + m^4}} + \frac{n^2a - n^2v}{n^2 - m^2}:$$

therefore, after striking out whatever cancels each other, dividing the whole numerator by n^2, and placing before the radical sign its root n, extracting the root of the denominator, and thence reducing the whole numerator to the same denominator as given by the author

$$z = \frac{n^2a - n^2v + n\sqrt{m^2a^2 - 2n^2av + n^2v^2}}{n^2 - m^2}.$$

The root must yet be extracted of the part which is under the radical sign by infinite series, and that will be of this sort.

$$ma - \frac{n^2}{m}v + \left(-\frac{n^4}{2m^3} + \frac{n^2}{2m}\right)\frac{v^2}{a} + \left(-\frac{n^6}{2m^5} + \frac{n^4}{2m^3}\right)\frac{v^3}{a^2}$$
$$+ \left(-\frac{7n^{10}}{8m^9} + \frac{10n^8}{8m^7} - \frac{3n^6}{8m^5}\right)\frac{v^5}{a^4}$$
$$+ \left(-\frac{21n^{12}}{16m^{11}} + \frac{35n^{10}}{16m^9} - \frac{15n^8}{16m^7} + \frac{n^6}{16m^5}\right)\frac{v^6}{a^5}$$

That n which is in front of the radical sign, should be injected into each numerator of this series; for it belongs to the entire series; then let them be added to the first two terms, $+ n^2a$, and $- n^2v$ themselves, which are connected with the radical; and each term be divided by the common denominator $n^2 - m^2$, thus will have the first [term] $n^2a + nma/n^2 - m^2 = na/n - m$; for the second $(-n^3v/m - n^2v)/(n^2 - m^2) = -n^3v - n^2mv/n^2m - m^3 = -n^2v/nm - m^2$; for the third $(-n^5/2m^3) - (n^3/2m)/n^2 - m^2$; to rearrange it properly, m^2 must be introduced into the numerator of the second part so that the common denominator of both parts be $2m^3$; and multiply by it the common denominator $n^2 - m^2$ whence this whole member will become $- n^5 + n^3m^2v^2/2m^3n^2 - 2m^5a$; which divided in the usual way will give the quotient $n^3v^2/2m^3a$,. And in the same manner proceeding through the entire series we shall finally have this:

$$\frac{na}{n - m} - \frac{n^2v}{mn - m^2} - \frac{n^3v^2}{2m^3a} - \frac{n^5v^3}{2m^5a^2} - \frac{5n^7 - n^5m^2v^4}{8m^7a^3}$$
$$- \frac{7n^9 - 3n^7m^2v^5}{8m^9a^4} - \frac{21n^{11} - 14n^9m^2 + n^7m^4v^6}{16m^{11}a^5}.$$

XVIII.

I derived this series here at length, not because some use of it will follow in such details, but to see from the analysis itself more clearly what has to be said now about it. For I have said in §III above, that, as it is given by the illustrious author, it labors under a double defect, one of them is this. From what he wrote after the fourth term in which the series ends in his case, anyone would rightly expect it to continue according to the same exact rule, namely, that the next term would be $n^7v^4/2m^7a^3$ which, however, is far from the truth, and so in the rest. The other of these [defects] is that the author should have noted that these signs which by him and also above are given, serve only when the angle of the refracted ray is greater than the angle of the incident one, whence R or n is greater than I or m: i.e., when the light rays go out of a denser medium into a rarer: when, however, contrariwise, all the signs after the second must be changed, whence for these rays the series will be $ma/m - n - n^2v/m^2 - mn + m^3v^2/2m^3a + n^5v^3/2m^5a^2$ etc., and after the first term is taken away, the aberration will remain as $= n^2v/m^2 - mn - n^3v^2/2m^3a - n^5v^3/2m^5a^2$ etc. For it is evident from figs. 13 and 14, that in this case the aberration always diminishes in fig. 13, in the other, however, increases. And if in special cases numerical values are used in the Huygenian way, namely for $I:R$, or $m:n$, 2:3, or else 3:2, according as the circumstance requires; entirely the same numbers will come out in both cases and exactly the twins of those which we have derived from the Huygenian figure in §§XIV and XV.

XIX.

Before this discussion about the rays parallel to the axis is dismissed, when I am about to take up the demonstration of Huygens's rules, it is just that something should be said also of those numerical values which he has given that serve by a greater profit in the investigation of errors and of their finding: the first ones of them which occur are (page)[29] 9/2 of the thickness, that is of the versed sine of the convexity of the glass, for the error of the rays coming out from the denser medium into the rarer, and (page)[30] 4/3 of the same, for those entering the denser from air; and also (page) for the entire plano-convex lens, whose convex side is turned toward the light rays, 7/6 of the same thickness, in the same way also for the equally convex lens (page)[31] 5/3 of the thickness, that is 5/3 of the versed sine of convexity. Since that distinguished gentleman was of an admirable sagacity, it seems to me, that he brought out all these no other way than by testing the matter sometimes through trigonometry or some other way, which has happened to me also in some [cases] while working on problems of this nature, before I started a more accurate investigation through calculus, or before I inspected these rules. But that these numbers are not at all accurate is abundantly clear from the precedents, where it is evident that 9/2 and 4/3 are only the first numbers

which originate from the Newtonian series according to the different values of I and R, or m and n, and likewise in the rest.

XX.

But it could not be judged as to be easy to explain by what art or by what kind of research he derived the rule which is given (page)[32] for the investigation of the aberration of a lens with non-equal convexities; where supposedly the semi-diameter of the convexity facing the rays equals a, the semi-diameter of the other equals n, the thickness of the lens equals q, he stated the error [to be] equal to

$$\frac{27a^2 + 6an + 7n^2}{6 \times a + n^2}q$$

(for it ought to be written thusly): it is not put, however, in such an obscure [way] that we perhaps could not follow it. It is quite certain to me how it happened to be in these first numbers 27 in the numerator $= m^3$, and 6 in the denominator equals mn; and since in all such trials, as the experts know, the testing should be by the extreme [cases]: now one of these extremes is the lens equally convex on both sides, whose aberration was found to be 5/3, and the other, the plano-convex lens, whose error is 7/6. But since it might happen that when one semi-diameter is unity, that is $= 1$, the other rises even to ten thousand times as much or more, the numbers should be tempered so that if both a and n were unity, the sum of which $= 2$, the square is 4, $6 \times 4 = 24$ (it will be clear soon why the six should be used here), then the numerator must be, for equal semi-diameters, 40 (namely to make $5/3 = 40/24$). Because of the other extreme (namely, the plano-convex lens, whose error is 7/6) it was necessary indeed that a [factor of] 7 be adjoined to n^2, and also that a six be in the denominator: which will show more clearly from an example. Let, therefore, a and n be identical, and both $= 1$, and for the already given reason; namely that the denominator $6 \times \overline{a + n^2} = 24$, the numbers in the numerator no matter how they are set up have to make 40 together, to be $= 5/3$, and since each of these a^2, an, n^2 are $= 1$, it would be just the same if we would put $38a^2 + an + mn$, or $a^2 + 10an^2 + 29n^2$, or any other way; as long as the sum of the coefficients does not exceed 40. But if we put $a = 1$, but $n = 1000$, $27a^2$ will be $= 27$, $6an = 6000$, $7n^2 = 7,000,000$, whose sum equals 7006027 and $6 \times \overline{a + n^2} = 6012006$, which two properly placed are not far away from 7/6.[33]

---

[29] Huygens, *Œuvres Complètes* **13**, 1: p. 285.

[30] *Ibid.*, p. 287.

[31] *Ibid.*, p. 291.

---

[32] *Ibid.*, p. 293. Huygens used the equation

$$\frac{27aaq + 6anq + 7nnq}{6(a + n)^2} = \text{ED.}$$

[33] The conclusions reached in this sentence are Logan's.

XXI.

From the same equation (page),[34] however, he evolved another one for the investigation of the error if the semi-diameter a of the outside surface, and the focus of the lens which he puts $= d$, are given, whence since according to §XVI in the previous part of this [paper], $d = 2an/a + n$ follows by legitimate reduction [that] $n = ad/2a - d$, which substituted properly in place of n everywhere in the previous rule $(27a^2 + 6an + 7n^2/6 \times \overline{a + n^2})q$, aberration will come out $= (27a^2 - 24ad + 7d^2/6a^2)q$; which granted that it looks different from the first, but since it is deduced from it and rests on the same foundation, has to be judged the same entirely.

XXII.

All these having been disposed of, we should now see about the aberrations of the light rays that are incident obliquely on spherical surfaces, which, as it was said before, no one, as far as I know, has calculated yet, nor studied. It will be proper to proceed in their investigation the following way.

Let in fig. 15 the convex spherical surface be BN, whose semi-diameter $CB = CN$, axis AX; let AN be a light ray falling obliquely on it at N, the right sine of the arc BN is NV, the versed sine BV. By §X of the first part, if we extend AN to D, and draw from the center C its normal CD, after dividing it in the ratio of refraction, through the point of division we draw the arc dEf, and construct to this the tangent NEX, meeting the axis at point X, this point X will be the true focus of the oblique ray incident at N and hence refracted, whose aberration XZ from point Z, the primary focus of the same surface (which is to be found by §X) could be investigated by one of the following methods.

XXIII.

The illustrious Newton said, as it was mentioned before, "In the same way as the errors are determined for the parallel incident [rays], similar ones can be determined for the divergents or convergents (i.e. obliques) though by a more difficult calculation." But the whole difficulty of this calculation consists in the first place in this that instead of a parallel ray, whose length being infinite has no ratio, here we have to consider not only the distance of point A, whence the rays come forth or toward which they converge, from the vertex B; but, if we must insist on higher accuracy also the very length of the ray AN; in the second place also (what is of greater importance) in this, that the ratio of refraction should not be taken

[34] Huygens, *Œuvres Complètes* 13, 1: p. 295. Huygens used the equation $DE = \dfrac{27aaq + 24adq + 7ddq}{6aa}$, DE being the aberration.

from the right sine NV at all, but from the straight [line] CD; what will be evident abundantly to [those who] examine properly §XXIII and whatever follows in the first part.

XXIV.

Since, indeed, in what is already discussed, there are shown various methods to determine the errors of the rays parallel to the axis but [which are] incident farther from it, the same here through whichever of them [methods] be; taking into account only the light ray AN, in this manner, we could examine further, whence the variety, that originates therefrom.

XXV.

Let's put therefore the length of the ray AN fig. 15 $(= \sqrt{AV^2 + VN^2}) = a$, $CB = r$, $AC = c$, $NV = d$, the versed sine BV ([that] of the right [sine] NV) $= v$, the versed sine EG (of the right one CE) $= y$, whence follows $CV = r - v$, $NE = r - y$, and let's call $CX = x$. But since the triangles AVN and ADC are similar, it follows that $AN : NV :: AC : CD$. i.e., $a : d :: c : dc/a$, hence $CE = dnc/ma$: and also $NV : VX :: CE : EX$, i.e., $d : x + r - v :: \overline{ndc/ma} : nc/ma \times \overline{x + r - v} + r - y = NX$. But also $CE : CX :: VN : NX$, i.e., $\overline{ndc/ma} : x :: d : max/nc$: hence the equation $nc/ma \times \overline{x + r - v} + r - y = max/mc$; if s is substituted in it for ma/c, it will be $sx/n = n/s \times \overline{x + r - v} + r - y$, from which [if] properly rearranged as in §VII, it will come out $x = nr/s - n - n^2v + sny/s^2 - n^2$, the last part of which is the true aberration, and in the same way as it was told in that section, it could be factored into these parts $(n/2)v + y/s - n \pm nv \sim y/2s + n$. Further something noteworthy occurs in the first part, namely, if we put $a =$ the distance of the point A from the vertex of the lens B, since $s = ma/c$, this $nr/s - m$ will express the true distance of the primary focus to be found according to §V from the center C.

XXVI.

But this problem could be solved by a somewhat easier method (as it would seem to some perhaps) without substitution and using only the letters m and n: namely, if in fig. 16 the straight line $CE = ndc/ma$ of the previous paragraph and figure is drawn not from the center C but from the point N, and after the arc dEf is constructed with the same distance as before, the tangent CES is drawn to it from the center C: for now in the triangle CSD they made up, everything will be as in the triangle NXV in figs. 11 and 12 in §IV, for here SN will be in the ratio of refraction to SC, as there was XC to XN, or $n : m$ whence putting $CN = r$ and $NS = x$, through an entirely similar calculation to that which is given in §§V, VI, we shall have x itself; assuming, namely, that the versed sines are in duplicate ratio to their right [sines], that is, $x = (n/m - n)r - (n^2/m^2 - mn)v$ or if greater accuracy is needed and y itself is to be used as in §VII, x will be $= (n/m - n)r - n^2v + mny/m^2 - n^2$, whose last part could be factored, as it was already said often. Having been found this way the value of this $x = NS$ and CS constructed, let's draw NX parallel to it, thence, two entirely similar triangles will emerge ASC and ANX, but if SN is taken from AN, it follows that $AS: SN::AC:CX$, whence we shall have the same point X as before. For it makes no difference whether the same distance is erected from C or let down from N; what is, however, absolutely necessary is that the straight lines be parallel to each other.

XXVII

But if the method by series pleases someone more, he will find in the illustrious Newton's "Arithmetica Universalis" problems to find the foci of divergent rays, what in the first edition of that excellent work, in the year 1707, is problem 17. But in the second, problem 31. From which a series could be derived to this purpose, though with a labor exceeding the worth of a city [i.e., it would involve more labor than it would be worth] if with less it could be made up some other way; which, indeed, will be clear very easily from taking notice of the preceding [discussions]. For let us call in the preceding fig. 15, as before $AN = a$, $AC = c$, $CN = CB = r$, $BV = v$, $CX = x$, but not let NV be $= \sqrt{2rv - v^2}$, and because of similar triangles ANV, ACD, follows $CD = c/a\sqrt{2rv - v^2}$, and therefore $CE = nc/ma\sqrt{2rv - v^2}$. And, also, as before, VX will be $= x - r - v$, and its square $= x^2 + r^2 + v^2 + 2rx - 2vx - 2rv$, to which be added the square of $NV = 2rv - v^2$, whence we shall have $\sqrt{x^2 + r^2 - 2rx - 2vx} = NX$, again because of similar triangles NVX, CXE; it follows that

$$NX: NV::CX: CE = \frac{x\sqrt{2rv - v^2}}{\sqrt{x^2 + r^2 + 2rx - 2vx}}.$$

But it was found before [that]

$$CE = \frac{nc\sqrt{2rv - v^2}}{ma},$$

whence arises an equation, after elimination of its numerators the radical $\sqrt{2rv - v^2}$ common to both, [there] will remain

$$\frac{x}{\sqrt{x^2 + r^2 + 2rx - 2vx}} = \frac{nc}{ma},$$

and by cross multiplication

$$max = nc\sqrt{x^2 + r^2 + 2rx - 2vx},$$

or $(ma/nc)x = \sqrt{x^2 + r^2 + 2rx - 2vx}$, let's call as before $ma/c = S$, and squaring the whole it will be

$(s^2/n^2)x^2 = x^2 + r^2 + 2rx - 2vx$, and properly rearranged as it was shown in §§XLVI and XLVII in an entirely similar case, it will give

$$x = \frac{n^2r - n^2v + n\sqrt{s^2n^2 - 2n^2rv + n^2v^2}}{s^2 - n^2}.$$

from here we can proceed in the same way as in §XVII, this series will come out entirely similar to that except that here instead of $m s = ma/c$ is used, and that here we discuss the rays entering the denser medium, namely

$$x = \frac{s}{s - n}r - \frac{n^2v}{s^2 - ns} - \frac{n^3v^2}{2s^3r} - \frac{n^5v^3}{2s^5r^2} - \frac{5n^7 - n^5s^2}{8s^7r^3}v^4$$

etc. It should be noticed further, that in all these are given methods not only to investigate the errors of refractions, but also the very absolute foci of the rays no matter how far from the axis they fall, if we compare these foci with those whose investigation is given in the first part of this [paper], we can derive therefrom the value of the aberration easily.

XXVIII.

Further, in exactly the same way as the focus of the oblique ray AN is investigated, the focus of NX itself (taking in account the different media), also has to be investigated, whence the focus of the entire lens will be given; which if put together with the focus found by the Huygenian rules demonstrated in the first part, then the entire error of any ray, in any lens is very easily detected.

III. MATHEMATICS, PHYSICS, AND NEWTON

The publication of Logan's two treatises on mathematical optics established him as one of the leading amateur mathematicians in North America.[1] Good amateur mathematicians in the colonies were not plentiful, to be sure; simple arithmetic was not stressed in the grammar schools, and students were admitted to the colleges without knowing the barest essentials of arithmetic. The better mathematicians in Logan's America were the professionals, notably Isaac Greenwood and John Winthrop IV of Harvard College.

---

[1] Tolles, *James Logan and the Culture of Provincial America*, p. 205. The best colonial mathematicians were the professionals, notably Isaac Greenwood, whose *Arithmetick Vulgar and Decimal* was published in Boston in 1729, and Dr. John Winthrop. Both Greenwood and Winthrop taught at Harvard College. James Winthrop, a librarian at Harvard and the son of Dr. John Winthrop, was active in mathematical study, although he was not as learned a mathematician as his father. Cadwallader Colden, whose training was in the field of medicine, dabbled in mathematics, but with indifferent results. (Brooke Hindle, *The Pursuit of Science in Revolutionary America, 1753–1789*, pp. 46, 93; Florian Cajori, *The Early Mathematical Sciences in North and South America* [Boston, Richard G. Badger, Publisher, Gorham Press, c 1928], p. 21.)

Still there were amateur mathematicians, and, although none of them were up to the standard of a Newton, Leibniz, or Euler, not even of Greenwood or Winthrop; nevertheless some were not bad. Their mathematics was not sophisticated; it did not extend much beyond plane geometry, algebra, and trigonometry; but some amateurs used what they had imaginatively. Witness Cadwallader Colden, even though his attempted solution of the cause of gravitation was not only erroneous, but confused. Witness also James Logan who used what mathematics he had not only imaginatively, but with meticulous accuracy, and without confusion.

As a mathematician, Logan was entirely self-taught. In his youth in Edinburgh, Logan had studied and mastered William Leybourn's *Cursus mathematicus*, which contained most of the mathematical and astronomical knowledge of the time. His private library at Stenton contained the ancient Greek mathematicians and such modern mathematical works as the books of John Wallis. He procured from England a copy of Sir Isaac Newton's *Principia mathematica* in April, 1708, and he was reputedly the first colonist in English America to own a copy of that significant work. Logan mastered Newton's system of calculus, and purchased the two subsequent editions of the *Principia* as they came from the press. His admiration of Newton is well known; he followed the trend of eighteenth-century Anglo-American thought in accepting Newton as master. However, as the sequel will show, Logan was never a slavish follower of Newton. When he thought the master was in error, he did not hesitate to say so, and endeavored to prove him in error.

An example of his critical treatment of Newton is his attempt to improve upon Newton's formulas on the elliptic arc. His interest in Newton's formulas was occasioned by his search, in 1736, for some rule by which he could calculate the length of an elliptic curve.[2] The only method of solving the problem that he could find was by the doctrine of infinite series with fluxions.

Logan had been aware of Newton's fluxions, but his busy life had not previously allowed him time to study them. Now with the problem of the elliptic arc before him, he devoted his time to a close study of fluxions. Yet, in his manipulation of infinite series no working knowledge of fluxions is apparent. Logan employed Newton's algebraic notations, played around with Newton's formulas, and discovered that one of Newton's series did not converge.[3] Thomas Godfrey then worked on the problem and encountered

the same difficulty. Logan concluded therefrom that Newton's method was in error, and he developed a different method which proved more successful. Logan appended his method as a postscript to his letter to William Jones, dated June 14, 1736.

Jones's replies to Logan's letters do not appear to be extant,[4] but Logan's letter of July 25, 1737, indicates that Jones, in a letter dated February 20, 1737, had attempted to excuse Newton's mistakes on the grounds that the mathematical philosopher had only offered some illustrations by way of examples, and the illustrations tended to oversimplify because the purpose was to render the problem understandable to general readers. Logan rejected that excuse. It seemed to him that Newton had deliberately concealed his errors by offering misleading illustrations. For Logan it was a matter of honor that the mathematician offer only what is "well grounded and perfectly true." He acknowledged the genius of Newton, but he doubted that knowledge of Newton's fluxional methods was little more than "a pretty amusement," unless one had the genius and "knack at invention" to make practical use of them.[5]

In Logan's time, of course, calculus was not yet fully developed in England and America. After 1830 Newton's fluxional method was replaced by modern differential and integral calculus—an advance over both Newton's and Leibniz's pioneering work. It may be, as one historian suggested in 1942, that Logan, in his formula for the convergence of the infinite series, worked out mathematical canons which may have some value for modern mathematicians,[6] but it does not appear that it has had any influence on mathematical thought. Nevertheless, Logan's speculations indicate the direction of the mathematical thought of a colonial virtuoso before the middle of the eighteenth century.

[2] According to Brasch, the problem of determining the length of an elliptic curve, "has the characteristic properties of problems in orbital motion." Frederick E. Brasch, "James Logan, a Colonial Mathematical Scholar," *Proc. Amer. Philos. Soc.* **86** (1942): p. 9.

[3] See Logan to William Jones, Pensilvania, 14 June, 1736, on p. 64 *infra*.

[4] Brasch, "James Logan, a Colonial Mathematical Scholar," *loc. cit.*, p. 5.

[5] Logan's knowledge of Newton's method of infinite series appears to have been based on his reading of Newton's letter to Henry Oldenburg, June 13, 1676, which was published in the *Commercium Epistolicum* (1712). The letter may be found in *The Correspondence of Isaac Newton*, edited by H. W. Turnbull (2 v., Cambridge, University Press, 1960) **2**: pp. 20–47, 53–55. There were errors in Newton's illustrations of the infinite series in his 1676 letter. *Ibid.*, **2**: p. 46.) It does not appear that Logan was aware, at least at the time his letter to Jones was written, of Newton's *The Method of Fluxions and Infinite Series . . . from the Author's Latin Original not yet made public*, translated and edited by John Colson (London, 1736), but he owned a copy of Newton's *Opticks* (1704) which contains as an appendix the *Curvatura curvarum*, the first publication of Newton's fluxions.

[6] Frederick E. Brasch, "James Logan, a Colonial Mathematical Scholar," *loc. cit.*, p. 9. For a good brief discussion of the development of calculus through the seventeenth century see A. R. Hall, *The Scientific Revolution*, 1500–1800: *The Formation of the Modern Scientific Attitude* (Boston, Beacon Press, paperback, 1956), pp. 230–233. A more complete study is Carl B. Boyer, *The Concepts of the Calculus: A Critical and Historical Discussion of the Derivative and the Integral* (New York, Columbia University Press, 1939), especially pp. 187–233.

Logan's attention was again distracted by other concerns, and, thinking himself too old anyway, he resolved not to attempt any further mathematical studies. But, however he may have felt about his age (he was sixty-three in 1737), he had too curious and energetic a mind for any long period of intellectual inactivity, and for him, mathematics was not as much labor as it was an amusement. During the winter of 1737–1738, he studied a problem which had been put by David Gregory in the first Latin editon of his *Catoptricae et dioptricae sphaericae elementa*.[7] He thought Gregory's solution a good one, but resolved to try another procedure. In a letter to William Jones he reported that he had worked out a biquadratic equation from fluxions, and had attempted to use Edmund Halley's method of resolving the biquadratic equation. However, he had discovered errors in Halley's rules, and had dropped the subject for a time. The mathematics was difficult, time-consuming, and once again, Logan thought himself too old to undertake such tasks.

In March, 1738, however, Logan's interest in the problem was revived when he read in Du Hamel's *History of the Academy of Sciences* that as late as 1699 European scholars had been attempting, without success, to solve it. After some reflection a solution occurred to him, which he hastened to send to William Jones in England.

The problem of the duplication of the cube had puzzled ancient Greek mathematicians. According to ancient Greek tradition, a tragic poet, who evidently did not know mathematics, had represented Minos as putting up a tomb to Glaucus. Minos, however, was dissatisfied with the size of the tomb, and the poet represented him as specifying that the tomb be doubled in size by increasing each of the dimensions in that ratio. Geometers took up the question and attempted to work out a method of doubling a given solid while keeping the same shape. Because geometers used the cube, the problem took the name of duplication of the cube. Hippocrates of Chios found that by "finding two mean proportionals in continued proportion between two straight lines the greater of which is double of the less, the cube will be doubled. . . ."[8] The problem now was to find the two mean proportionals, and Hippocrates offered no solution.

Ancient Greek tradition has it that certain Delians were commanded by the oracle to double a certain altar without altering its shape, and they struggled unsuccessfully with what now became known as the Delian problem. The problem was referred to geometers who were with Plato in the Academy, and solutions were attributed to Eudoxus, Menaechmus, and erroneously to Plato. Logan evidently referred to the solution attributed by Eutocius to Plato.[9]

Upon reviewing the ancient authors, Logan realized that his offered solution was hardly new, and he wrote to William Jones at once to correct his error. He asked that, if Jones showed Logan's solution to the Delian problem to anyone, he should show with it Logan's letter of correction so that he would not be thought presumptuous.

The letter of correction (May 4, 1738), however, also contained a criticism of Edmund Halley's "rules for resolving affected equations,"[10] which had been published in the *Philosophical Transactions* and republished elsewhere. No evidence has come to light that Halley ever saw or learned of this letter, and one can only speculate as to what his reaction may have been if he had read it.

Logan's interest in mathematics and Newtonian physics led him to a study of musicological problems. In the medieval quadrivium musicology ranked with arithmetic, geometry, and astronomy as a learned discipline, and it interested Logan because of its relationship to mathematics and astronomy. As his letters of January 17 and 22, 1716/17 show, his mind ranged from tuning stringed instruments to the measurement of the diameter of the earth, and the two problems did not seem as unrelated to him as they might seem to us. The musicological problem was the precise measurement of sound waves issuing from musical strings in motion. The study of this problem caused Logan to wonder whether musical strings and pendulums did not obey the same laws of motion.[11] The speculation was interesting, and it could not be proved, as Logan knew, without difficulty and costly experiments simultaneously on various places on the earth—experiments comparable to those which had been undertaken with pendulums by the Royal Academy of France in the past. Logan concluded,

---

[7] David Gregory, *Catoptricae et dioptricae sphaericae elementa* (Oxford, 1695). Gregory (1661–1708) was the Savilian Professor of Astronomy at Oxford from 1691 to his death. His *Catoptricae et dioptricae sphaerica elementa* was written for undergraduate students in college and contained the substance of lectures he had delivered at Edinburgh in 1684. The treatise was reprinted at Edinburgh in 1713, translated into English by Sir William Browne in 1715, and a second edition of the English translation was published in London in 1735. See *Dictionary of National Biography* 8: pp. 536–537.

[8] George Sarton, *A History of Science: Hellenistic Science and Culture in the Last Three Centuries B. C.* (Cambridge, Harvard University Press, 1959), p. 110.

[9] Sir Thomas Heath, *A History of Greek Mathematics* (2 v., Oxford, Clarendon Press, 1921) 2: pp. 244–246, 255.

[10] "Affected equations" are better known as "affected or complete quadratic equations." "Equations containing the square or second power of the unknown quantity but no higher power are called *quadratic equations*. A *pure quadratic* contains only the square; an *affected* or *complete quadratic* contains both the square and the first power. The equation $25x^2 + 18 = 3x^2 - 8$ is a pure quadratic; $50x^2 - 5x = 125$ is a complete or affected quadratic." (Edward E. Grazda, ed., *Handbook of Applied Mathematics* (4th ed., Princeton, N. J., D. Van Nostrand & Co., Inc., 1966), p. 107.

[11] Logan's intuition was correct. See Brook Taylor, "De Inventione Centri Oscillationis," Royal Society of London, *Phil. Trans.* 28 (1713): pp. 11–21.

therefore, that his speculation would never be tested by experiment. Logan's speculation was, of course, without scientific foundation, as Governor Hunter appears to have advised him, and Logan abandoned the speculation, but not without regret.

In his letter of January 17, 1716/17, Logan alludes to a "phonometer." The allusion is interesting, because as late as 1823 the invention of a "musical Phonometer" was still a matter of speculation.[12] Yet Logan refers to the phonometer as a "discovery." Whether the word "discovery" means invention of the instrument or the discovery of the idea of such an instrument is not clear. However that may be, Logan saw in the phonometer not only an instrument for the visual measurement of sound waves but an instrument which might prove useful in determining the diameter of the earth.

European scientists had long attempted to determine the diameter of the earth by means of pendulum experiments in various places on the globe. Newton based his determinations of terrestrial gravity and diameter on pendulum experiments which had been conducted by the Royal Academy of France. Logan's suggestion was that a more sophisticated method than pendulum gravity determinations might be worked out. His attempt to discover a relationship between acoustics and such determinations appears to have been original, and it never amounted to more than a speculation. Nevertheless, it arose from a detailed knowledge and understanding of physical and mathematical problems which occupied the attention of many European scientists during the period and which remain important problems in our own day.

The mathematical and physical problems which attracted Logan's attention and aroused his imagination arose chiefly from his study of Newton, although he was inspired also by his reading of other authors such as Mersenne and du Hamel. Newton was the central figure, and like most of his educated Anglo-American contemporaries Logan paid due homage to the English mathematical philosopher whose shadow fell across the eighteenth century as Einstein's shadow has fallen across the twentieth. Logan's admiration for Newton the scientist was qualified, however, by occasional criticism of Newton the man.

Logan was easily antagonized by any appearance of injustice, falsehood, or misrepresentation in any man, including Newton. Hence, he took exception to the portrait of Newton which appeared in the third edition of the *Principia Mathematica*, because it showed the author as he was thirty or forty years earlier. More seriously, he noted that John Flamsteed's name had been frequently expunged from the third edition, evidently because of the serious disagreement which had arisen between Newton and Flamsteed in their old age.

What disgusted Logan most, however, was the omission of the Scholium from Book II, Proposition 7. In that Scholium Newton had credited Leibniz with the independent invention of a differential calculus similar to Newton's. The Scholium had appeared in the first two editions of the *Principia*, but it was dropped from the third edition, and a new Scholium, which contained no mention of Leibniz, was substituted for it. The omission from the *Principia* of any mention of Leibniz's contributions to the development of differential calculus marked the climax of a quarter-century of controversy, and it appeared to Logan that it was dishonorable.

Newton and Leibniz, working independently, had discovered a method of calculating by fluxions, i.e., the differential calculus. Twentieth-century scholars are in general agreement that the two mathematicians had arrived at similar conclusions during the 1660's and 1670's. However, Leibniz published his method earlier than Newton, first in a letter to Henry Oldenburg in 1677, then in the *Acta Eruditorum* in 1684. Before the appearance of the first edition of the *Principia Mathematica* in 1687, Newton seemed reluctant to publish a formal exposition of his method, although his *Analysis per aequationes*, which Leibniz saw in 1676, contained applications of it. His *Method of Fluxions*, written in 1671, was not published until 1736. Newton first published his fluxions, however, in the *Curvatura Curvarum*, an appendix to his *Opticks* of 1704. Until 1699 Leibniz was not challenged as an inventor of differential calculus; Newton, in the first edition of the *Principia*, had not raised the question of priority of discovery. In 1699, however, Nicolas Fatio de Duillier (1664–1753) published *Lineae brevissimi descensus investigatio geometrica duplex*, a book in which de Duillier expressed the belief that Newton was the first inventor of the differential calculus and that Leibniz had plagiarized Newton's work. De Duillier, a Swiss who had settled in England and whose book was published by the Royal Society in London, based his conclusions on the fact that in 1676 Leibniz had examined the papers of John Collins in London and had seen papers by Newton which contained applications of the new method. Florian Cajori, historian of mathematics and editor of a twentieth-century edition of Newton's *Principia*, has written of this incident:

During the week spent in London, he [Leibniz] took note of whatever interested him among the letters and papers of Collins. His memoranda discovered by C. J. Gerhardt in 1849 in the Hanover library fill two sheets. The one bearing on our question is headed "Excerpta ex tractatu Newtoni Msc. de Analysi per aequationes numero terminorum infinitas."

It seemed to Cajori that on the basis of the memoranda Leibniz "seems to have gained nothing pertaining to the infinitesimal calculus."[13]

[12] *New Monthly Magazine* 8 (1823): p. 20, as quoted in *Oxford English Dictionary*, 1933 ed., 7: p. 789.

[13] Florian Cajori, *A History of Mathematics* (2nd ed. rev., New York, Macmillan, 1961), p. 214.

The publication of de Duillier's *Lineae brevissimi* triggered a century of charges and countercharges. Repeated charges that Leibniz had plagiarized Newton's method in 1676 and Leibniz's appeal to Newton and the Royal Society of London for justice spurred the Royal Society to appoint a committee to investigate the matter. The committee's report, *Commercium epistolicum*, was first published in 1712 and was republished in 1722 and 1725. Its final conclusion was that Newton was "the first inventor." The committee did not find that Leibniz had plagiarized Newton's method, but suggested that he might have. The omission from the third edition of Newton's *Principia* of any mention of Leibniz as an independent inventor of differential calculus followed. The long-range consequence was the disruption of all communication between English and continental mathematicians during the eighteenth century. Because Leibniz's notations in differential and integral calculus were superior to Newton's, the progress of mathematics in England was retarded thereby.

Logan gave Leibniz credit for priority of publication which he believed more important than priority of invention. He was inclined to assign the latter to Newton, but he strongly suspected that the entire controversy had been "blown up" by powerful members of the Royal Society who opposed what Logan called "the house that had so long employ'd Leibnitz" and which appears to have been the House of Hanover. As H. G. Alexander's *The Leibniz-Clarke Correspondence* (1956) shows, the controversy over priority spread after 1710 into an angry dispute between Leibniz and the Newtonians as to the validity of much of Newton's philosophy.

The whole Leibniz-Newton controversy embarrassed Logan. To Logan it was a "fierce unnatural Dispute," and he wished that both Newton and Queen Anne (a stronger monarch might have put a stop to it) had been "gathered to their ancestors" before 1710. There can hardly be any question that Logan admired Newton, but it appeared to him that Newton's fame had been tarnished by his involvement in an ill-tempered dispute with the German philosopher during the sunset years of his life.

LOGAN TO WILLIAM JONES, PENSILVANIA, 14TH OF JUNE, 1736 RIGAUD, pp. 288–292.

My good friend,

Though I have no absolute right to claim so much of thy regard, as to believe thee from thence under an obligation to answer my letters; yet I have had such proofs of it in other respects, that I really thought I might at least hope for the favour, and I still so far continue of the same opinion, that I here venture once more to solicit it on an occasion, that is of some importance to a subject I have now in hand; and my reason for troubling thee with it is, because I know

none more capable of answering my doubts, or of whom, notwithstanding thy past silence, I can more freely request such a favour.

The case is this. My former busy life having never allowed me to look so deeply into the mathematics as otherwise I might, I entirely declined considering the doctrine of infinite series with fluxions till lately, when being desirous to find some certain method or rule for knowing the length of an elliptic curve, and meeting with no other than that of series by fluxions, I resolved to try this, and meeting with those two in Sir I. Newton's letter of the 13th June 1676, published in the Commercium, 8 vo. edit. p. 137, 138, the first for finding one part by the absciss from the centre, and the other for that taken from the vertex,[14] I doubted not but I was master, or soon should be so, of what I sought. But, vastly to my disappointment, I found in working on them, that the last, which should give the most important part of the curvature, is no converging series at all; but after a few steps in the quotient, unless x be taken very small, (and then it is of no use,) it actually diverges, if I may use the expression; I mean that the differences increase and grow wider.

Surprised at this, and being diffident of myself, I got T. Godfrey, who has more youth on his side, and an excellent natural genius for such studies, to try what he could make of it; but he found it just as I did. As this could not but raise my wonder, I applied myself more closely to consider the process, and there discovered the reason, viz. that the series from which this root (in p. 138.) is extracted, has its second, or some other near term, bigger than the first, (it is the unit or 1, in the second, third, or fourth term, which exceeds the first,) and therefore it must be impossible in nature ever to deduce any good series from it.

Being disappointed in this, I tried what might be done for the same part by throwing out x and working on y; and this succeeded greatly to my satisfaction; for I found another series, by which I could continue this last on y as far as I pleased, without extraction, (a point I laboured for in the other on x, but could never find it.) I also struck out a canon, by which I could with much greater expedition and certainty carry on the operation in numbers; and from these steps I doubted not but I was master, though with more labour than I expected, of all I had proposed by it; and yet in the close I found it all wholly lost, otherwise than that I had the satisfaction to find that what I sought for, at least by any of these methods of series hitherto delivered in books, was utterly impracticable to any considerable exactness; that is, I take it to be impossible by the method of series and fluxions, as applied by the first inventor's, or other subsequent author's directions, to find the length of one whole quadrant of an ellipsis, whose axes are 2, 1,

---

[14] Both are in Example 5 in Newton to Henry Oldenburg, June 13, 1676. See fn. 36 below.

to ten places. Wolfius, in his second edition of his Elements, (p. 463), so lately as 1730,[15] bestowed no less than four pages on shewing the process of an operation on the absciss from the centre, but to no manner of purpose; and since he has shewn no other method for those rectifications, but thought fit, so very lately, to enlarge that chapter above what it was in the first impression, I conclude that he knew no other, or at least none that he thought better: and whether there by any other at all known, (for what they call evoluta I think will do nothing,) is more than I am able to discover. Trying the parabola I found it worse, and so did T. G.; yet at length, from a close application to the subject, he discovered one fluxional equation of it to be the same with that of the logarithmic; and by substitution of this found means for giving the length of the parabolic line, which by any direct process in the way of series appears impracticable.[16]

Upon the whole, I much question whether it is possible, even by approximation, to rectify any of the conic curves besides the circle, (which by its uniformity in all its parts becomes more easily manageable,) unless by accident;[17] nor am I yet satisfied whether 'tis possible in nature to find a right line equal to any curve whatever.

Therefore as these are my doubts, and in this situation I am entirely out of the way of conferring with the more skilful on these heads, I request, in behalf of thy own art, and, as I may in some sense say, thy profession, to think seriously of the subject, and after a proper inquiry and examination of what is here offered, which I am apt to think will be new, and perhaps somewhat surprising both to thyself and other proficients in geometry, to favour me with thy thoughts on the subject, as soon as may possibly suit thy conveniency, lest, if I am in error, it may prove of ill consequence in another affair, wherein I am led to mention this. I have added, below, the equations I proceeded on in working on Newton's second series, which I carried to a considerable length before I discovered its imperfection.

. . . .

thy assured friend,
J. Logan

The equation for the ellipse is $rx - exx = yy$; x from the vertex And I find

$$\dot{z}\dot{z} = \dot{x}\dot{x} \times \frac{rr - 4rex + 4e^2x^2 + 4rx - 4exx}{4rx - 4exx}$$

---

[15] Christian Wolff (1679–1754), *Christiani Wolfii . . . Elementa matheseos universal.*

[16] To the twentieth-century mathematician this statement would appear to be naïvely expressed, although correct. The parabolic arc is rectifiable by means of logarithmic functions.

[17] Here Logan contradicts what he has just said about the parabolic arc. I am indebted to Dirk J. Struik for this observation and for the observation in fn. 16.

Dividing this last part I make the quotient $r/4x + 1 - 3/4e + eex/4r + e^3x^2/4r^2 + e^4x^3/4r^3$, &c. which, after Newton's example, I multiply by rz, to make the first term a square, and then extract, which gives the same that stands in the book, drawing in again $\sqrt{rx}$, dividing by rx, &c., &c.

My series for the arch from the vertex on y.

Equation of the ellipse $rx - exx = yy$; $r = $ lat. rectum, $e = r/$Transv. D.[18]

$$y + \frac{2}{3r^2}\ \left.\begin{matrix}y^3 + 8e \\ - \frac{2}{5r^4}\end{matrix}\right\}y^5\ \left.\begin{matrix}+ 32e^2 \\ - 16e \\ + \frac{4}{7r^6}\end{matrix}\right\}y^7$$

$$\left.\begin{matrix}+ 128e^3 \\ - 96e^2 \\ + 48e \\ - 10\end{matrix}\right\}y^9\ \left.\begin{matrix}+ 512e^4 \\ - 512e^3 \\ + 384e^2 \\ - 160e \\ + 28\end{matrix}\right\}y^{11}$$

&c. &c.

LOGAN TO JONES, PHILADELPHIA, JULY 25, 1737; RIGAUD, pp. 313–319.

My good friend,

I am favoured with thine of the 20th of February, but cannot say whether more to my satisfaction on the one hand, or to my confusion on the other; to the first it was greatly, nor could it be less to the other, to observe what trouble, without any just claim on my part to the favour, I had given thee. I have, however, by it the most pregnant instances of thy goodness, thy reflection on which may also yield a sincere satisfaction to thyself, as nothing but a large share of benevolence, the highest of human virtues, could have led thee to such a condescension. For I cannot omit remarking that to have shown thy method for rectifying that curve (the ellipse) would have been abundantly sufficient, without proceeding to the operation itself, which I perceive from its slow convergency must have been unreasonably operose to carry it even to ten figures: for I find, after a few steps, the decrease in the terms of the series will, for that ratio of diameters (2:1), be but little more than 1/4 in each, and to come at those ten figures the series must, if I mistake not, be carried to above eighty places. My obligation is therefore the greater. But the fineness of the invention, the neatness of the demonstration, with the exactness and plainness of the process, still add highly to the pleasure; and, upon the whole, I must with sincere gratitude ever acknowledge myself very much thy debtor.

From what I touched on in these mathematical subjects, I suppose it was not difficult for thee to take a pretty exact measure of my small skill in them. 'Tis what I never pretended to; nor did I ever look

---

[18] $\dfrac{r}{\text{Trans. D.}}$ means $\dfrac{r}{\text{transversal diameter}} = \dfrac{r}{\text{major axis}}$.

into them, otherwise than now and then for my diversion, and because I would not be wholly ignorant of those sciences, that I was assured, on all hands, were both highly useful and entertaining; and therefore, though without the instruction of any master, I resolved to get some little acquaintance with them. But my life has been generally a course of business, and the hours I could borrow from that have been employed in a continued variety, I cannot say of study, but of amusement from books promiscuously; and it was but by accident that I looked into the method of series and fluxions applied to them, which I had avoided through an apprehension of their being too intricate: but endeavouring (as I think I formerly hinted) to find the length of an elliptic curve, the proportion of which to that of the circle, considering their sines are exactly in the same ratio and differ only by the prolongation of one axis, I thought might easily be discovered; but trying several methods for it, and still finding myself disappointed, I resolved to try it by that method of series which I thought must infallibly give it me, and accordingly, with some little application, made myself master of those given by Newton in his letter of July, 1676, drawn up for Leibnitz.

His first series for x from the centre proved easy, and much the more so, because he gave another for continuing it; his other for x from the vertex, as he had found himself obliged to use much more contrivance in it, gave me more trouble, and still the more, because I could not find any possible method for continuing it without actually working it. By that means, however, I struck out a canon of my own, which exceedingly facilitated that and all other such operations. Having gained this, and carried the series a good many steps further, I then concluded I was absolutely master of the whole; but when I came to reduce it to numbers, behold! I was just as far off my point as ever; the series after a very few steps actually diverged and ran still wider. I then threw away my whole work with indignation; and with no small amazement to find it possible for Sir Isaac Newton to commit such a blunder. I examined, however, my own work, as I also did his, over and over, and at length I discovered what he had overlooked, the true source of the error, as I mentioned it to thee (with more, I think, of what I am writing) in a former letter; that is, that the second or third term in the series, whose root was extracting, was greater in value than the preceding; a circumstance that, if

---

[19] This equation is from Newton's letter to Henry Oldenburg, June 13, 1676. The vinculum bent into a vertical stroke symbolized aggregation. The comma also symbolized aggregation, and it appears to have been added by Logan. For Newton's notation for negative and fractional exponents see Florian Cajori, *A History of Mathematical Notations* (2 v., Chicago, Open Court Publishing Co., 1928) **1**: pp. 217, 354–355. Parentheses to indicate aggregation were rarely used until recent times. See *ibid.* **1**: p. 284.

I mistake not, will destroy the convergency in any extraction whatever. But of this, I observed, that great man was not at all sensible at the time he was writing that letter, for in his very second example of extraction, which is to find $\overline{c^5 + c^4x - x^5},$ $|^{1/5}$ [19] he tells us we may use $- x^5$ for P, as well as c^5, but that the first is preferable if x be very small (*valde parvum*), and the latter, if it be very great, when the truth is, that if x be ever so little greater than c, (and to make them equal would be absurd,) the series produced would not converge at all.

And hereupon I must crave thy pardon if I cannot admit of thy excuse for such mistakes, as if they were only set, as sums are to young people, for their practice in multiplying or dividing; for in that letter the author first very clearly and fully shewed his method of extracting, and freely communicated that most excellent and admirable invention of the uncial series, as I call it, but he appears to have designedly concealed, under pretense of avoiding tediousness, his method of applying fluxions to series, choosing only to give some illustrious examples of what might be performed by them. It therefore directly concerned his honour, that all he offered in that way should be perfectly well grounded and exactly true. Yet notwithstanding all this, though it shews that no one man can see every thing, and that Sir I. Newton could mistake, the world, as long as there remains in it any regard for science and sound knowledge, must ever revere that wonderful man's memory, and acknowledge him the greatest genius, in that way, that has ever been known to this day. But I must add, that though the knowledge of these methods is a pretty amusement, yet without a genius and extensive capacity, and particularly some knack at invention, they appear to me of but little use to be learned or studied. And thus far I have run on, chiefly to excuse the past trouble I have given thee: the appearance of a mistake in Sir Isaac Newton I thought was worth inquiring into, and by the favour of thy answer I am fully satisfied.

· · · · · · · · · · · · · · · · · ·

LOGAN TO JONES, STENTON, IN PENNSYLVANIA, MARCH 31, 1738; RIGAUD, pp. 323–326.

My esteemed friend,

Last year I returned thee my thanks most justly due for thy very obliging letter; and now send this, not to give thee, as my former did, any trouble; but rather for our mutual diversion, if it should happily yield thee any, as it gives me some in writing, though the subject I own is of no great importance; but the following is what at present offers.

It is now about two years since I almost resolved to look no further into any mathematical subject; yet it has once or twice happened otherwise. For this last winter making some inquiries into that of light, which I carried to some length, and by I know not what accident casting my eye on D. Gregory's problem, page 48 of his own first Latin edition of his Catopt. et Dioptr. Elementa, I could not think his solution just, and therefore I proposed the problem to myself in another manner, which having duly wrought out by fluxions, I had a biquadratic equation, for resolving whereof I had recourse to Dr. Halley's method, first published in the Transactions, and since subjoined to the other treatises. But on proving his rules, which I examined with some care, I found two or three considerable mistakes in them, and thereupon began an attempt to set that most necessary part of algebra in a clearer light; since to form an equation, the principal business of that art, is of no use without knowing how to solve it when formed. But I found it was no proper time of life with me to attend such studies, and therefore dropped it.

But occasionally looking the other night into Du Hamel's History of the Academy of Sciences, (ad annum 1699,)[20] on a very different view, I happened in his 558p. to read these words: Doctor Ramondus Coninkius praepositus palatii Limae Sacello libellum edidit, Limae excusum anno 1696, in quo solutionem famosi problematis de cubi duplicatione a se inventam putat. Hunc librum misit ad Academiam D. Bruyn-steem urbis Brugensis quaestor, atque ejus ea de re sententiam rogavit. D. de la Hire, cui id muneris datum est, ut problematis solutionem expenderet, paralogismum in ea dilitescentem et satis involutum deprehendit. [Doctor Raymond Coninkius, director of the chapel of the palace of Lima, has published a booklet, printed in Lima in the year 1696, in which he thinks that he has found the solution of the famous problem of doubling the cube. D. Bruynsteem, the treasurer of the city of Bruges, sent this book to the Academy, and asked their opinion about this issue. D. de la Hire, to whom it was given as a task to investigate the solution of the problem, detected a paralogism hidden in it and a quite involved one.][21] Now that so much notice should, so lately as in the year 1699, be taken of this problem, which consists only in finding two mean proportionals, I thought somewhat strange, and reflecting on it the next day, (yesterday), a method for it occurred to me, so very simple and easy, that I could scarce think it possible but it must have been hit on by somebody long before. I hereupon

examined what we have on the subject from Eutocius, in his Comment on Archimedes, L. 2. de Sph. et Cyl.; from Pappus's 3d and 4th books; from Eratosthenes, published in Greek by Bp. Fell, at the end of Aratus; also what Philander on Vitruvius has on the same, who is particular on it, page 178, &c. of Laet's edition; but I could nowhere find any thing resembling mine.[22] And the excellent Dr. Barrow, in his Compend. odit. of Archimedes, (p. 31.) on the place before quoted, having these words, Hoc problema solidum est, ad ipsius quippe solutionem exigens, ut duae mediae proportionales inveniantur, quod praestare nequit geometria communis, regula tantum utens et circino; per conicas sectiones, et aliis compluribus modis effici potest de quibus hic taceo [This problem is a solid one, for it requires to its solution that two mean proportionals be found, which common geometry, using only ruler and compass can not offer. It can be done by conic sections and by several other ways of which I keep silent here.][23] I from thence concluded that, in his time at least, there was no plain and simple method known for it. Therefore as I take mine to be truly such, and the problem in itself is allowed to be a noble one, if there are any such left, as I hope there are divers, who have yet a regard for geometrical effection or construction, and depend not wholly on logarithms and series, such a solution may be acceptable to them, if not known before. If it be, it is but little labour lost, and I have my invention only to myself; if otherwise, I leave it wholly to thy discretion and judgment to communicate, or, if proper, to publish it, in which view I have drawn it up in Latin, as

---

[20] Jean Baptiste du Hamel (1624–1706), *Regiae scientiarum academiae historia, in qua praeter ipsius academiae originem & progressus, variasque dissertationes & observationes per triginta quatuor annos factas, quam plurima experimenta & inventa, cum physica, tum mathematica in certum ordinem digirentur.* 2 ed. priori longe auctior. Autore Joanne-Baptista du Hamel . . . (Parisiis, J. B. Delespine, 1701). The first edition appeared in Paris in 1698.

[21] Philippe de Lahire (1640–1718) was the author of a number of books "*dans le style des anciens.*" Ferdinand Hoefer, *Histoire des mathematiques depuis leurs origines jusqu' au du dix-neuvieme siecle* (Paris, 1886). De Lahire was the author of *Sectiones conicae in novem libros distributae* (Paris, 1685) and of *Nouveaux Elements des Sections Coniques* (1679).

[22] Eutocius of Ascalon (on the Palestinian coast) flourished in the first part of the sixth century B.C. His *Commentarii in Opera Non Nulla Archimedis*, including Book 2 on "The Sphere and Cylinder," was printed in *Archimedis Opera Non Nulla a Federico Commandino Urbinate Nuper in Latinem Conversa, et Commentariis Illustrata* . . . (Venetiis, apud Paulum Manutium, Aldi F., 1558). Pappos of Alexandria flourished in the second part of the third century B.C. Pappos's commentary on Archimedes was incompletely translated into Latin by Federigo Commandino (1509–1577) and published at Pesaro in 1588 (Ferdinand Hoefer, *op. cit.*, p. 261). It also appeared in a volume of the Princeps of Aristarchos with Commandino's Latin translation and Pappos's commentary, by John Wallis (1616–1703) (Oxford, Sheldonian Theatre, 1688). Ἀράτου Σολέωϛ φαινόμενα καὶ διοσημεία. *Accesserunt annotationes in Erastosthenem et Hymnos Dionysii.* (Oxonii, E. Theatro Sheldoniano, 1672), 2 parts in 1 v., was edited by J. Fell, Bishop of Oxford. Aratos of Soloi (ca. 315–213?) was the author of two extant astronomical poems: *Phainomena and Diosemeia.* Marcus Vitruvius Pollio was a Roman architect who flourished about 10 A.D. The edition of his work cited by Logan is *M. Vitruvii Pollionis de architectura libri decem. Cum notis . . . Philandri . . . Barbari . . . Salmasii . . . Elementa architecturae Collecta ab . . . Henrico Wottono . . . Lexicon Vitruvianum Bernardini Babi . . . et ejusdem Scamilli . . . Vitruviani. De pictura libri tres. Leonis Baptistae de Albertis. De sculptura excerpta . . . ex dialogo Pomponii Gaurici . . . Ludovico Demontiosii Commentarius de sculptura et pictura . . . Omnia collecta . . . & illustrata a Joanne de Laet . . .* (Amstelodami, L. Elzevit, 1649).

[23] Isaac Barrow, *Archimedes opera: methodo nova illustrata, & succincte demonstrata* (Londini, G. Godbid, 1675).

I find most things of the kind are in the Transactions, or the benefit of foreigners as well as our own people.

When I began this, I expected more time, but now find myself reduced to nearly the last moment I can have by this opportunity, and therefore must abruptly close, which I do, with sincere respect and gratitude, from

thy affectionate and obliged friend,
J. Logan

Quantum se torserint veteres de problemate Deliaco (ut olim nuncupatum est) expediendo, seu cubi duplicatione, i.e. duobus mediis proportionalibus inveniendis, ex Eutocio, Eratosthene, Pappo et aliis satis notum est. Huic enim unico problemati deberi fertur conchoidis et cissoidis inventio, quinimo primos in conicis progressus hinc ortum duxisse creditur. Solutiones ejus diversimodae repertae sunt, at per-

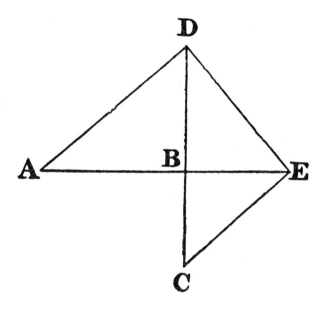

plexae vel operosae fere omnes, ut ex dictis autoribus abunde patet. Quid melius et paratius ad effectionem geometricam, (nam de numeris constat,) a nuperioribus praestitum sit nescio. Mihi vero occurrit solvendi methodus, quae si nondum aliis innotuerit, haud indigna fortasse videbitur quae publici fiat juris.

Problema solidum esse, et simplici geometria, seu recta linea et circulo solvi non posse, a geometris perhibetur. At si sequens methodus, sola rectarum ope, perinde facilis comperiatur atque Euclideorum problematum paene dixerim facillimi solutio, quidni pro pure geometrica haberi poterit vix constare videbitur. Ea autem haec est.

Datae duae linae AB et BC ad angulum rectum in B componantur, producantur autem versus E et D; ducantur porro duae parallelae per data puncta A et C, secantes alteras productas in D et E, ita vero, ut linea DE has parallelas connectens sit utrique perpendicularis; et factum est quod quaeritur, ut ex

ipso intuitu patet. Triangula enim CBE, EBD et DBA, propter angulos recto sunt similia, ideoque ut data BC ad BE, ita BE ad BD, et ita BD ad datam AB, et inter datas duae sunt mediae BE et BD. Q.E.D.

Regeretur fortasse effectionem non pure geometricam esse; non directe enim dari puncta E et D, sed experiendo tantum reperiri. Sed cum norma aut quadra instrumenta sunt aeque ac circinus geometriae simplici inservientia, hujus ope res facillime effecta dabitur: promptissime vero, si adhibeatur scala illa apertilis seu diductilis, quae in thecis organorum geometricorum ad ducendas parallelas plerumque habetur. Si alterum latus enim ad punctum A applicetur, et alterum ad C, duo autem anguli ejus ad rectas BD et BE pariter applicentur, quod factu est facillimum, expeditissime definientur haec puncta. Atque hoc pacto duo media geometrica, haud operosius quam unicum ex Euclidis praecepto, dantur.

I.L.

[It is well known from Eutocius, Eratosthenes, Pappus and others how much the ancients tortured themselves to solve the Delian problem (as it was once called) or the doubling of the cube, i.e., to find two mean proportionals. For the invention of the conchoid and cissoid is due, it is said, to this one problem, even it is believed that the first progress in the conics originated from here. There were found different solutions of it but almost all perplexing or laborious, as it is evident abundantly from the mentioned authors. I don't know if anything better and handier was offered by more recent ones to make a geometrical solution (for the numerical is clear). But a method of solution occurs to me, which if it is not known to others yet, perhaps, will not seem unworthy to be published.

That the problem is a solid one, and cannot be solved by simple geometry, namely, by straight lines and a circle, is told by the geometers. But if the following method, using only straight lines, should be found altogether easy and a solution of—I would say—almost the easiest of the Euclidean problems, why can it not be regarded as a purely geometrical [one] we hardly can see. It is now the following:

Let two given lines AB and BC be constructed at right angle at B, and continued toward E and D; let further two parallels be drawn through the given points A and C, intersecting the other two in D and E, but in such a way, that line DE connecting these parallels be perpendicular to both; and is done what is required as it is clear from just looking [at it]. For the triangles CBE, EBD and DBA, are similar because of the right angles, and therefore as the given BC to BE, so BE to BD, and so BD to the given AB, and between the given two BE and BD are the means. Q.E.D.

It is objected perhaps that the construction is not purely geometrical because the points E and D are not given directly, but are found only through experimenting. But since the right angle or square [drawing] instruments serve the simple geometry just as the compass, with their help the matter will be most easily settled: most promptly, indeed, if we use a scale which can be opened or pulled apart, which is found commonly in the cases of geometrical tools to draw parallels. For if one side is applied to point A, and the other to C, the two angles, however, are applied at the same time to the two lines BD and BE, which is very easily done. These points will be defined very quickly. And this way the two geometrical means are given, with no more effort than the one from Euclid's rule.]

LOGAN TO JONES, PHILADELPHIA, 4TH OF MAY, 1738; RIGAUD, pp. 327–33.

My esteemed friend,

This comes solely to crave thy excuse for my last, wherein I ventured on a solution of the Deliac problem; and though what I gave will, I imagine, be found the readiest and best method of any that have been applied for geometrical construction, yet having since discovered what from my first thought of it I suspected, that it is now new, I must own myself much too rash in sending it away; but, an opportunity just then presenting, I unhappily made use of it, and the article, in what I said both in my letter and the other paper, that gives me the most concern, is my mentioning those ancients, Eratosthenes, Pappus, and Eutocius, as having left us no such method as mine in their writings, which I now find is a mistake, and was thus occasioned.

Eratosthenes and Pappus I have, and thought I also had in Rivaltus' (alias Flurantius's) edition of Archimedes, in folio, Paris 1615, all that Eutocius had left in his comment on that great author; but looking some time after into G. Vossius de Scient. Math., I there found he had in that comment given the several methods of twelve ancient mathematicians, all named there [p. 72] by Vossius.[24] Hereupon recollecting that I had seen, in this country, amongst some books that had belonged to a learned German, who many years since was coming over hither, but died in the way, a translation of Archimedes's works into high Dutch by Sturmius, in a pretty large folio, I procured it; and there found all those several methods described at large, the very first of which, being Plato's, proves exactly the same with mine.[25] But as he directed the making of an instrument for the purpose, of which the figure is there given, it appears the more mechanical; yet in effect that instrument and the other I mentioned, which we have in our instrument-cases for drawing parallel lines, are very much the same. Thus by that defect in Rivaltus, of which I was not then apprehensive, (for I never yet could meet with Archimedes with Eutocius on him, either in the Greek edition of Basil, or of Commandinus's Latin version,)[26]

I was led into the mistake. And I now earnestly request that if any other persons of skill in these things, besides thyself, should have seen my letter or inclosed paper before this comes to hand, thou wouldst be pleased also to shew him this, for removing the imputation of a presumption, of which I would by no means be guilty, and therefore not be thought so.

But having further mentioned Dr. Halley's rules for resolving affected equations, first printed in the Philosophical Transactions, and several times since in other books, as if there were some errors in them; lest I should be thought to mistake in this also, I shall here, from my notes when on the subject, point out some, though very briefly, referring to the tract itself to render what I say more intelligible.

For the unknown letter or root he substitutes, as is done in all the methods directed for the same purpose, a binomial $a \pm e$, and valuing a by a number, the nearest that can be guessed at for the root, with that number $+$ or $- e$ he brings out another equation, which gives the value of e only, and to find this from the first three terms of it, viz. the absolute number, for which he puts b, and e and ee with their coefficients, expressed by s and t that is from $\pm b$, $\pm se$, $\pm tee$, he forms a quadratic equation, and solves it in the common way, by completing the square, though he explains not his process to his readers.

But owning there is some difficulty in judging whether e be $+$ or $-$, he gives these two rules: that the signs of b and s must and will always be contrary, and that if we have $+ b$, we must put $- se$, and therefore 'tis $a - e$ or a is taken too great, and on the contrary $- b$ gives $+ e$, and then a is taken less than just.

But these rules, especially the latter, do not at all hold, and one cannot sufficiently admire, that so great a master in those sciences, as that gentleman is known to be, should so widely mistake in a point of the greatest moment in his rules, and which he might easily have seen into with a little application. For the sign of b, $+$ or $-$, depends not so much on taking a too great or too little, as on the proportion of the

[24] *Archimedes opera quae extant. Novis demonstrationibus commentariisque illustrata per Davidem Rivaltum a Fluriantum . . .* (Parisius, apud Claudium Morellum, 1615). For Vossius see Part Two, fn. 16.

[25] Logan's reference was to Johan Christoph Sturm (1635–1703), a German mathematician and physicist, who translated the works of Archimedes into German. Logan owned *Unvergleichlichen Archimedis Kunst-Bucher* (Nuremberg, 1670) and *Des Unvergleichlichen Archimedis Sand-Rechnung* (Nuremberg, 1667); both books bound together.

[26] ΑΡΧΙΜΗΔΟΤΣ, ΤΟΥ ΣΤΡΑΚΟΥΣΟΥ, *Archimedis Syracusani Philosophi ac Geometrae Excellentisimi Opera, quae quidem extant, omnia, multis iam seculis desiderata, atq a quam paucissimis hactenus uisa, nuncq primum & Graece & Latine in lucem edita . . .* (Basileae, Ioannes Heruagius excudi fecit, 1544). The Basle edition contains Eutocios's commentaries on Archimedes in Greek. For Commandinos's Latin translation see footnote 22 *supra*; this work contains the Latin translation of Eutocios's commentaries on Archimedes. Logan did not own a copy of Commandinos's translation, but he had the Basle edition.

coefficients to each other, in their respective places, which ought naturally to bear some proportion to a gradation of the roots ascending, which will best appear from an example.

The fourth power of $x - 3 = 0$ is $x^4 - 12x^3 + 54x^2 - 108x + 81 = 0$. Let some of these coefficients be varied, still keeping up the same value;

(A) $x^4 - 8x^3 + 54x^2 - 144x + 81 = 0$
(B) $x^4 - 16x^3 + 54x^2 - 72x + 81 = 0$

or all of them thus;

(C) $x^4 - 10x^3 + 51x^2 - 126x + 108 = 0$
(D) $x^4 - 14x^3 + 58x^2 - 90x + 45 = 0$

Then if we try all these by $2 + e = x$, we shall have these remainders:

from

	b	s	t	u
(A)	-39	$+8e$	$+30e^2$	$* + e^4$
(B)	$+41$	$-16e$	$-18e^2$	$-8e^3 + e^4$
(C)	-4	$-10e$	$+15e^2$	$-2e^3 + e^4$
(D)	$+1$	$+6e$	$-2e^2$	$-6e^3 + e^4$

Again if we try them by $4 - e = x$, we shall have these:

from

	b	s	t	u
(A)	$+113$	$-160e$	$+54e^2$	$-8e^3 + e^4$
(B)	-111	$+152e$	$-42e^2$	$* + e^4$
(C)	$+36$	$-58e$	$+27e^2$	$-6e^3 + e^4$
(D)	-27	$+42e$	$-14e^2$	$-2e^3 + e^4$

Now as all these in the first range downwards arise from $a + e$, and all in the second from $a - e$, yet we have $+ b$ and $- b$ indifferently in both, we see his principal rule does not at all hold. And in (C) and (D) of the first range, we have both $- b - se$, and $+ b + se$, which also contradicts the other.

These examples might be sufficient to prove what I have said; yet I shall further observe, that when an equation has more real or affirmative roots, to say nothing of others, these rules must often necessarily fail; for instance, this equation $x^4 - 38x^3 + 529x^2 - 3192x + 7056 = 0$ has these two roots $x = 7$ and $x = 12$, and, if tried by 10, it gives the remainder $+ 36 \pm 12e - 11e^2 \pm e^3 + e^4$: it gives also the very same if tried by 9; and whether we take $+ e$ or $- e$, b is always $+ 36$, and the value of e is $= + 2, - 2, + 3,$ and -3, that the numbers may answer both roots, as $9 + 3 = 12, 10 - 3 = 7; 10 + 2 = 12$ and $9 - 2 = 7$.

So this, $x^4 - 140x^3 + 7324x^2 - 169680x + 1468800 = 0$, has these four roots 30,34,36,40, and if tried by

           b t

35, it gives the remainder $25 * - 26e^2 * + e^4$, in which $e = + 1, - 1, + 5,$ and $- 5$; and if tried by 32 and by 38, it gives the same for both, viz. $- 128 \pm 48e + 28e^2$

$\pm 12e^3 + e^4$, in which e is $= + 2, - 2, + 4, - 4, + 8,$ and $- 8$, that it may equally answer all the several roots; a speculation truly delightful to see, how admirably numbers thus qualified are formed.

Further again, if a true canonical equation or power be tried by any other numbers equally distant from its just root, the remainder for each will be the same: as if $x^4 - 24x^3 + 216x^2 - 864x + 1296 = 0$, the fourth power of $x - 6$, be tried by 4 and by 8, each of them will give the same, viz. $+ 16 - 32e + 24e^2 - 8e^3 + e^4$. But is it not strange, that if in that equation instead of $- 24x^3$ we put $- 25x^3$, the root will sink from 6 to less than 3,5 and if we put $- 23x^3$, it becomes, if I mistake not, impossible?

I shall yet add, that as all quadratics cannot be solved by completing the square, so it happens here that when the signs of b and t are alike, (the meaning of which is, when to form the equation, we must have $- b$, and the Doctor should have so explained it,) if b is greater than $1/4 ss$, the solution, by this method, becomes impossible, as in the preceding second (C) where $36 \times 27 >$ [is greater than] $(29)^2$, or which is the same $36/27 (= 1 1/3) >$ [is greater than] $(29/27)^2$; and so in many others. I am sensible, however, that in many cases this is an excellent method; yet there are some others that I would generally prefer.

I could add some other remarks, but even this from me may justly require an apology; for I own it is trifling away my time. But since I have for some years past been confined to crutches, amusements are necessary, and I fell into this solely by resolving one biquadratic equation (as I said) on the subject of light, which I was then closely considering. These remarks, however, may be the better excused, since the resolution of affected equations is an article of as great importance, as any I know, in algebra; for it can be of no use to know how to bring a question to an equation, unless we can resolve it when informed.

LOGAN TO ROBERT HUNTER, JANUARY 17, 1716/17, LETTER BOOKS OF JAMES LOGAN Vol. II, p. 150, HSP.

I am heartily sham'd of my Dulness, that after I had so lately wrote what I did out of Mersennius[27] concerning the Note of a String Strain[e]d w[i]th such a Weight, I should want to be putt in mind that the whole business of tuning Strings might be made to depend on weights, but thinking my Self press'd in time, for David Sent me word he must goe by Noon, I turn'd my thoughts to no other part of the thing than what had partly been the Subject of another

---

[27] Marin Mersenne (1588-1648), French theologian, mathematician, and philosopher, was the author of *L'Harmonie universelle, contenant la theorie et la pratique de la Musique* (1636-1637). This work is devoted mainly to music, but contains digressions on various aspects of mathematical science. For a brief biographical sketch of Mersenne see *Nouvelle Biographie Générale*, **35**: pp. 118-123.

Lett[e]r viz whether Musical Strings & Pendulums had the same laws of Motion. I must therefore beg thy excuse for the trouble I have given thee in making me see a thing that nothing but a great degree of Stupidity or thoughtlessness could have kept out of view.

As to the Phonometer[28] it Self I have seldom mett with any discovery that pleased me better, for I'm clearly of opinion, that Seeing a tune can be so divided into So many lesser notes & effects of different weights may be more Sensibly & easily discovered by those tones than by Pendulums[.] but nothing near the properties mentioned in thy Lett[e]r for the difference of the Semidiameters of the Earth at the Poles & Equator is Settled by Newton in the last adition [*sic*] of his Principles at 1/227, or they are in proportion as 228 to 229 whose Squares are 51984 to 52441 w[h]ich is under 100 to 101, So that between the Poles & the Equator there could not be above 7 of those intermediate Notes in difference. But however that proves, the Speculation (for I doubt it cannot goe much beyond that, further than to try experim[en]ts at remote places as the ffrench have done by Pendulums at Cayenne &c)[.][29] I say the Speculation appears to me to be very fine, & tho' it may never be made so Serviceable to the Tritons[30] nor can be much wanted for discovering the Latitude yet I should be well pleased to See how the Experim[en]t is to be inferred from that most delicate Sense the ear to the more certain in the eye, which is so favourably promised in the next, & therefore shall claim my Right.

Logan to Hunter, Philadelphia, January 22, 1716/17, Letter Books of James Logan, Vol. II, p. 151, HSP.

I was so highly pleased w[i]th thy last but one for Setting me to rights about the weight applied to the tension of strings, that I took not time to think what I should write in answer, till G. Willocks was very near going, and then found my self too much straitned, to which to my trouble, for I'm really Sorry for the disappointment in so pleasant a Speculation, I must impute that expressed in mine, where I Said, I was clearly of opinion, Seeing a tone may be divided into many notes, the effects of weights may be more Sensibly discovered by Musick than by Pendulums for I now doubt this will be found a mistake, tho' I heartily wish it were otherwise. I'm Sensible disappointm[en]ts in matters of Speculation will Sometimes give as great uneasiness as those of Some importance in affairs of life. I witness it my Self in this very one before us, but as Truth to all ingenuous minds is the greatest and most valuable discovery I'm p[er]suaded it cannot be altogether unacceptable.

On reading thy last I turned immediately to Du Hamel's History of the Royal Academy of Sciences from their first Institution to the end of 1700,[31] who I knew had inserted all their observations of Pendulums in Several parts of the world, and found five Several distinct observations but not at all consistent w[i]th themselves Some of them making the difference of the Lengths from that of Paris 2 lines in places to the Norther[n] of those Countries where it was found to be but one & a quarter. I then call'd to mind that Sr Is[aac] Newton discovered of it in his Principles, and turning to him from the 20th Prop. of his 3d Book w[hi]ch in his 2d Edition is a long one, Spent chiefly on that Subject of the different Weights of Bodies, in the Several parts of our Earth. In this he layes it down as a Rule that Weights are reciprocally as their distance from the Center and not as the Squares,[32] as I had computed, and that those Weights and the lengths of Pendules [pendulums?] are directly proportional, he hereupon examines the different observations communicated to the Academy, and reduces them to a Calculus, and also gives us a table of the different Lengths of Pendulums to every 5 degrees, from the Equator to the Pole, but from the Latitude of 40 to 50 for every Single degree, because there the Variation is the greatest, as whoever considers the Curvature of an Ellipse will easily find it must be. This table as far as it relates to Pendulums I have here copied, lest the Book should not be at hand, but if it be as I Suppose it may, to make this the more easy for computation I have calculated his numbers in Decimals taking 100,000 for the length of the Equator and So increasing.

Now by this it appears that the Weight of a Body at the Equator is to the Weight of the Same at the Pole only as 100000 to 100437 w[hi]ch is (as near as these numbers will answer), as 229 to 230 w[hi]ch numbers I should have used in my last instead of 228 to 229, but my memory on w[hi]ch alone I depended, happend as it frequently does, to deceive me, tho' I had occasion Some time agoe to consider the difference of the Earths diameters w[i]th some thought,

---

[28] "An instrument for measuring or automatically recording the number or force of sound-waves." (*Oxford English Dictionary*, 1933 ed., 7: p. 789.) "We should not be surprised to see this uncertainty brought, in time, under mathematical controul, by the invention of a musical Phonometer, to indicate the precise strength of sound." (*New Monthly Magazine* 8 (1823): p. 20, quoted in *ibid.*) Logan's "phonometer" appears to have been more than a century early.

[29] *Cf. infra* this page.

[30] Logan's allusion to "Tritons" is not clear, unless he was referring to tritones. A tritone is a musical interval of three whole steps.

[31] See fn. 20 *supra.*

[32] Newton wrote that "weights are . . . inversely as the distances of the bodies from the centre of the earth . . .. The weights in all . . . places round the whole surface of the earth are inversely as the distances of the places from the centre; and, therefore, on the hypothesis of the earth's being a spheroid, are given in proportion." (Sir Isaac Newton, *Philosophiae Naturalis Principia Mathematica*, Book III, Proposition XX.)

but a Variation in Newton's last Edition from his first, in these numbers, caused a jumble.[33]

I mistook further in my last w[i]th so small absurdity in allowing 8 of the Smaller notes for the whole difference between the Equator and the Pole, for now I think it appears that the whole will not take in the compass of a third part of a Single division. Yet however this proves, I must confess the Speculation it Self, tho' never (it seems) to be reduced to practice, has given and will still continue to give me no small pleasure.

As to the Centrifugal force being greatest at the Equator, and diminishing to nothing, as it must, under the Pole, I take that to be the very Same w[i]th the difference of Gravity in these two places & that the whole Centrifugal force under the Equator is 1/229th part of the natural Gravitation of bodies to the Earth and that this being taken quite off, under the Pole, and therefore so much weight there added, it makes the whole in that Scituation 230[.]

The invention of that Instrum[en]t to reduce it from the ear to the eye Shews also the Strength of its Author's thought. I took the hint of bringing two drinking glasses to the Same tone by help of Water from Crousaz[34] and endeavoured at the experim[en]t but my ear was not exact enough to give it the keeness it has mett w[i]th from a finer.

I hinted at an occasion I have had of considering the different diameters of the Earth w[hi]ch I believe I have formerly mentioned to thee, if not it was this. Considering the Earth as a Spheroid a Meridian must be either greatly or nearly an Ellipsis, and a horizontal line a Tangent to that Ellipsis, a p[er]pendicular to

w[hi]ch Tang[en]t can never coincide w[i]th the Center but under one of its Axes that is under the Equator or Poles, But all ponderous bodies are allow'd to Gravitate to the Center, then if so, a plum line must not be perpendicular to the horizon. I computed the difference and found that in our Latitude of 40 deg[rees] it made an angle of 14 min. & about a half. In short I was so possessed of the thought, that I made a large experim[en]t with a level fram'd on purpose of about 18 feet & a perpendicular of about 8 feet by w[hi]ch to my mortification I found the plum line was most exactly perpendicular to the level as I proved by changing the ends of the latter & putting that w[hi]ch first stood Northwest & South. This Soon brought me to reflect on my former want of thought in not considering that if the Rotation of the Earth threw up the parts of its body especially Water in Such a proportion tow[ar]ds the Equator, the Same must of necessity hold in a plum line that Swung by a thread, & therefore had the Same liberty as the other to tend, & must tend the Same way &, according to all reason in the Same proportion.

This entirely destroyed the very foundation of what I laboured for, by bringing the difference I was in quest of to a perfect equality, w[hi]ch is far from being the case in thy thought, for it really, I believe, has a Solid basis in nature, were there any ears fine enough to be sensible of it.

JAMES LOGAN TO COL. BURNET, PHILADELPHIA, 7 FEBRUARY 1726/7, LETTER BOOKS OF JAMES LOGAN, Vol. III, Section A, p. 60, HSP.

· · · · · · · · · · · · · · ·

I rec[ei]v[e]d this winter from W. Inny's of Lond[on] the 3d Edition & a very fair one to the Eye of Sir I. Newton's Principia published by the care of Dr. Pemberton & doubt not but one of the same has reach'd thy hands. As in the 2d Edition, that Gentleman, truly wonderfull in other respects, to shew the weakness of humane Nature, & the prevalency of the Passions even in the Greatest, was led either by his own or other peoples Resentm[en]ts to expunge honest fflamsted's name as often as they could doe it where it was used in the first Edition in the 3d Book of which it frequently occur'd[;] so now in this third they have done, what I doubt impartial men of Labor thought & solid Judgem[en]t who alone ought in such cases to be considered, will look upon as a yet greater Instance of the same Infirmity in dropping the Scholium to the 2 Lemma between the 7 or 8 Propp. of the 2d Book wherein Leibnitz was named & his Discovery of the differential Method was fully taken notice of, & substituted another—mentioning the Author's Letter to J. Collins in 1672 w[hi]ch I doubt will scarce give so hon[ora]ble an Idea of that Great Man.[35] Tis certain the world was obliged only to

---

[33] "On theoretical grounds, Newton found the diameter of the earth at the equator to be to the diameter from pole to pole as 230 to 229, on the supposition that the matter of the earth was all 'uniformly dense.' The compression of the earth is accordingly (230–229) ÷ 230, or 1/230. The latest experimental determination of the earth's compression, derived from pendulum observations on the intensity of gravity, yields the value 1/297.4." (Florian Cajori, "Appendix," Note 41 in *Sir Isaac Newton's Mathematical Principles of Natural Philosophy and His System of the World*, translated into English by Andrew Motte in 1729. The translations revised, and supplied with an historical and explanatory appendix, by Florian Cajori [Berkeley, University of California Press, 1947), p. 664]. Cajori cited *United States Coast and Geodetic Survey, Its Work, Methods and Organization*, Special Publication No. 23, Washington, 1928, p. 127, as authority for the value 1/297.4. According to a recent report of the Coast and Geodetic Survey, "plans are being made for a worldwide satellite triangulation project in cooperation with the National Aeronautics and Space Administration and the Department of Defense. The resulting triangulation scheme, consisting of about 36 stations uniformly spaced around the world, will tie together all geodetic datums, yield information regarding the size and shape of the earth, and provide an accurate geometric framework for future satellite observations for geophysical purposes." "Pendulum gravity determinations are being superseded by gravity meter observations," and the results should be more accurate. (*The Coast and Geodetic Survey: Its Products and Services*, publication 10-2, Washington, 1965, pp. 29, 30.)

[34] Jean Pierre de Crousaz, *Traité de Beau* (Amsterdam, 1715), pp. 237–238.

[35] *Cf. Sir Isaac Newton's Mathematical Principles of Natural Philosophy and His System of the World*, the English translation

Leibnitz for the Publication of that method who was so fair as to communicate it in a great measure to Oldenburg in 1677 when Sir Isaac was so carefull of concealing his that he involved it in his Letter 1676 in strange knotts of Lett[e]rs that all the art & skill of the universe could never Decipher as giving only the number of each Letter that ent[e]red his Short Proposition.[36] And yet foreigners have generally been so just as to pay all possible difference [deference?] to Sir Isaac as an Inventor tho' till his Publication of the Principia in 1687 they never had any thing of it from him. I have often indeed wished that Sir Isaac himself had never ent[e]red into the Dispute, but would if it must be disputed, have left it to others, for then the world would have been inclined to doe him more Justice, than now perhaps they will when he is considered as a party, which he has so warmly made himself.

But there is not less Humour shewn in his Picture in the ffront[37] much more like W Leybourn in his own hair at the age of 40 or 50 than Sir Isaac Newton at 83 and by all those who have seen him of late, as I did, bending so much under the Load of Years as that with some difficulty he mounted the Stairs of the Society's Room, that Youthfull Representation will I fear be considered rather as an object of Ridicule than Respect, & much sooner raise Pity than Esteem[.][38]

. . . .

JAMES LOGAN TO GOV. WM. BURNET, 10 MAY 1727, LETTER BOOKS OF JAMES LOGAN, VOL. III, SECTION A, p. 73, HSP.

Having about two hours since been favoured w[i]th thy most obliging Letter by my good ffr[ien]d A. Hamilton, tho' hindred by finishing mine by a vessel just now sailing for Britain till the last minute of the Post stay as he affirms with us. Yet finding the 3rd Edition of the Principia had not yet reached N[ew] York I resolved to Charge him w[i]th it for thy perusal & on Inspection I believe the Picture mentioned in my last was actually drawn for Sir Isaac at the age of 83 will appear to have something uncommon in the fancy of somebody who occasioned such a Draught & the setting of it there.

Scarce any man living has had a greater veneration for that surprizing genius formed for an admission into the Secrets of Nature unknown to the whole race of mankind before than myself, & thy observations on him are so exceedingly just th[a]t they have nothing further to be said on that head. He is, however great, but a man & when I last saw him in 1724 walking up Crane Court, & the stairs leading to the Society Room, where I also had the opportunity of viewing him for about two hours, he bent under his load of years exceeding unlike what they have represented him two years after as in body[.] it's but reasonable to expect a Declension elsewhere[;] so that his for his own honour as well as the Nations to w[hi]ch he has been a very great one, had he & Queen Anne both been gathered to their ancestors, by the year 1710 before or & before [sic] that fierce unnatural Dispute broke out between him & Leibnitz w[hi]ch I always believed was blown up by the Forces of the Society in opposition to the house that had so long employ'd Leibnitz[,] they might have sett in their Horizon as I formerly thought w[i]th a somewhat greater Glory. Tis only from this way of thinking I dropt what I did of him in . . . [my previous letters to you]. If I have not altogether thy concurrence my tenderness for his Reputation I hope will be easily excused. I verily believe Leibnitz had the first hints from Newton's Letter & others concerning his Inventions & that from thence that Great Genius w[hi]ch we find in no man else did, built his great Superstructures—but from all I can find in the Commercium[39] I no where perceive that Sir Isaac intended any mortal should discover his method of working or what we call the algorithm of his fluxions & his having afterwards given us nothing new beyond what Leibnitz had published of that kind in the Acta Eruditorum[40] in 1682 [1684] is not altogether so much to his advantage as might be wished in his favour. Thus I have insensibly run on beyond expectation on the subject & having here as I find expressed my whole Sentim[en]ts upon it shall never I believe have occasion to give thee any further trouble on that head.

. . . .

IV. BOTANY

James Logan as a scientist is best known for his work in the field of botany. His controlled experiments with maize, in which he arrived at an accurate and detailed description of the role of pollen and the

---

of Andrew Motte in 1729 as revised by Florian Cajori, pp. 655–656.

[36] Henry Oldenburg (1615?–1677), a natural philosopher and man of letters, was one of the first secretaries of the Royal Society of London. Logan's reference is to Newton to Oldenburg, Cambridge, June 13, 1676. This letter may be found with English translation in *The Correspondence of Sir Isaac Newton*, edited by H. W. Turnbull (3 v., Cambridge, Published for the Royal Society at the University press, 1960), **2**: pp. 20–41. Logan alluded to a letter from Leibniz to Oldenburg written in 1677. This appears to have been Leibniz to Oldenburg, 11 June 1677, *ibid.* **2**: pp. 212–225. Also see Leibniz to Oldenburg, 12 July 1677, *ibid.* **2**: pp. 231–234.

[37] The portrait of Newton appears to have been drawn within a few years of 1691, and represents him as he looked about the time that he wrote the *Principia*.

[38] Logan saw Sir Isaac Newton during the winter of 1723. (Frederick B. Tolles, *James Logan and the Culture of Provincial America*, pp. 129–130.)

[39] Gottfried Wilhelm von Leibniz (1646–1716), *Commercium epistolicum de varia re mathematica, inter . . . I. Newtonum . . . G. Leibnitium . . . et alias . . . una cum recensione praemissa . . . controversiae inter Leibnitium & Keillium de primo inventore methodi fluxionum, etc.* (1725).

[40] *Acta Eruditorum* (50 v., Leipzig, 1682–1713). This was a periodical publication.

functions of the sex organs of plants in the reproductive process, were an important contribution to the science of plant physiology. Logan also contributed to botanical science by encouraging John Bartram in his observations and classification of North American flora and by bringing Bartram into transatlantic communication with European botanists, notably Karl von Linné (Linnaeus).

One historian of early American science tells us that Logan's contribution to botanical knowledge "revealed his close relationship to the English natural history circle."[1] Logan, in common with other colonial gentry in his time, had an agrarian outlook. Even though his economic interests were in commerce and shipping, he possessed the average colonist's love for the soil. He relaxed from the tedium of business and provincial government by puttering in his garden at Stenton. He communicated with Linnaeus, Gronovius, Collinson, and other figures in the international natural history circle. Yet his interest in natural history should not distract one's attention from the fact that Logan was as much, if not more, interested in natural philosophy, especially astronomy and mathematics. Moreover, it should be borne in mind that Logan sought in natural phenomena a key to the riddle of the universe. His interest in the sexual generation of plants was aroused by the broader philosophical implications of the subject.

Logan approached plant physiology through his reading in the field of natural religion. In 1726 he obtained and read William Wollaston's The Religion of Nature Delineated, a book which he greatly admired.[2] While reading Wollaston, Logan was impressed by a suggestion that the male might receive its semen from the air or food. The implication of that suggestion intrigued him. Was it possible that the seeds of life were transmitted to the male from the surrounding environment? Logan determined to find out.

Upon perusing Richard Bradley's New Improvements of Planting and Gardening, Logan found an observation that plants have male as well as female seeds.[3] The idea was, of course, as new as Pliny, but it had not previously occurred to Logan. He was excited. If it could be proved that female plant seed was impregnated by semen that entered the plant from the external world, then a completely different view of the creation might be the result. What was true of the vegetable kingdom might also be true of the animal. The true seeds of life might exist somewhere else "before they are cloathed with a visible corporeity."

Logan's reading in Wollaston and Bradley occurred during the winter, when he was unable to attempt experiments in his garden. It is likely that he read further into the subject during those winter months, although he may have read some of the titles he mentions concurrently with the experiments. As Logan's papers on his botanical experiments show, he became well acquainted with the literature of the field. He read Malpighi, Millington, Grew, Morland, and it was probably by reading Miller's The Gardeners Dictionary that he learned of Geoffroy. Surprisingly, he did not mention Rudolph Jacob Camerarius whose experiments in plant reproduction logically preceded and provided a basis for Logan's experiments with maize.

In the spring of 1727 Logan began his experiments with maize. In each corner of his garden he planted a hill of maize. When the maize had grown to a mature height, he began his controlled experiment. He cut off the tassels of the maize on one of the hills. On other hills he opened the ends of the ears. From some of that maize he cut off all the silken filaments; from others he removed varying amounts of the silken filaments, carefully noting the quantity removed from each. He then tied up the ends of other heads with a fine muslin that would prevent the passage of the farina, but would not obstruct air, sun, or rain, and he tied the muslin loosely so as not to stunt vegetation. Then he waited and carefully noted the results. The results convinced Logan that maize was fertilized by windblown farina, the semen in the sexual generation of plants. Logan chose maize, for it was easier to experiment with under controlled conditions, but he strongly suspected that all flora reproduced itself in this manner.[4]

Logan's work in plant physiology carried forward the earlier contributions of Marcello Malpighi, Sir Thomas Millington, and Nehemiah Grew. Malpighi and Grew were cofounders of microscopic plant anatomy and the first to describe pollen grains, although it appears doubtful that either one clearly understood

[1] Brooke Hindle, The Pursuit of Science in Revolutionary America, 1735–1789 (Chapel Hill, University of North Carolina Press, 1956), p. 22.

[2] William Wollaston (1660–1724) was a deist whose The Religion of Nature Delineated, published in London in 1722, was enjoying considerable popularity at the time Logan discovered it. In 1677 Leeuwenhoek had discovered spermatozoa, and the idea subsequently became quite popular that the male semen contained the individual in the small and that the female ovum served only as the shelter and source of nourishment. It is possible that Logan was aware of Leeuwenhoek's preformation theory of the genesis of life, but he referred only to Wollaston in this connection. Cf. A. Wolf, A History of Science, Technology, and Philosophy in the 18th Century 2: p. 467. Logan owned a copy of Leeuwenhoek's Continuato Arcanorum Naturae . . . (Delft, Henry A. Kroonevelt, 1697), and it is preserved in the Library Company of Philadelphia.

[3] Richard Bradley (1688–1732), New Improvements of Planting and Gardening, Both Philosophical and Practical. Explaining the Motion of the Sap and Generation of Plants. Several editions of this work were published in London before 1727. Logan had the 1726 edition.

[4] The role of insects in pollination was discovered later in the eighteenth century.

the role of pollen in the sexual generation of plants.[5] Camerarius, in the late seventeenth century, experimented with flowers to determine how they reproduce themselves. His experiments are now generally accepted as having proved that pollen was essential to plant reproduction, but they were not so accepted in the early eighteenth century.[6]

Logan appears to have been the first to observe by controlled experiment wind pollination in the sexual generation of plants. His work attracted attention in Europe from the beginning. Linnaeus used Logan's *Experimenta et meletemata* in his *Disquisitio de sexu plantarum* (St. Petersburg, 1760).[7] Logan's experiments with Indian corn were cited in botanical treatises well into the nineteenth century.[8] In 1749 J. G. Gleditsch convincingly established the necessity of pollination in the sexual generation of plants. The later work of Joseph Gottlieb Kohlreuter, whose report on flower pollination was published in four parts between 1761 and 1766, and Konrad Sprengel's researches in cross pollination in 1812 contributed even more greatly to the sexual theory. There were other colonial intellectuals who experimented with the hybridization of Indian corn and the role of wind pollination in the sexual generation of plant life, but Logan's work was the most influential among European botanists.[9]

In November, 1727, Logan wrote of his experiments to Thomas Goldney, a Quaker botanist in Bristol. Urging Goldney to undertake similar experiments, not only with maize but with other vegetables, Logan suggested that his findings would lead to a philosophical inquiry as to the origin of life and the nature of the creative process. To be sure, this was a speculation that might be beyond finite human understanding. Nevertheless, one could glimpse in such experiments something of the divine power by which the whole of creation was conducted. That this philosophical speculation was foremost in Logan's mind during the fall of 1727 is confirmed by his remarks in his letter to Alexander Arscott, dated November 29, 1727.

When Logan wrote to Goldney, it was his intention to visit England the following summer and join the English botanist in the proposed experiments. However, Logan changed his mind during the late fall of 1727 and winter of 1727–28 because of the fear of war in Europe. Then in January, 1728, he was crippled by an accidental fall, and that ended whatever hope he might still have had of visiting England. During the spring of 1728 he experimented with various flowers and the blossoms of trees, and used a microscope in examining the parts of flowers, but he did not achieve the success that he had enjoyed in his experiments with maize.[10]

During the next few years Logan was occupied with other affairs, and there is no evidence that he continued his botanical experiments after 1728. However, he probably continued his reading on the subject. In a letter to his brother, Dr. William Logan of Bristol, England, dated April 8, 1728, James Logan asked for a book entitled *Botanick Essays*, written by Patrick Blair and published in 1720. It may have been some time between 1728 and 1735 that Logan read of Geoffroy, a seventeenth-century French botanist who had experimented with maize. Geoffroy had reached the conclusion that seeds may grow to maturity without being impregnated by the farina. It appears to have been this observation by Geoffroy that moved Logan to write to Peter Collinson, the London Quaker merchant, of his own experiments which had proved that farina was essential to the sexual generation of maize. This letter, dated November 20, 1735, was subsequently published in the *Philosophical Transactions* of the Royal Society of London.

---

[5] According to Wodehouse, Grew "stated somewhat vaguely that he believed that fertilization was one of the several functions of pollen." (R. P. Wodehouse, *Pollen Grains: Their Structure, Identification and Significance in Science and Medicine* [New York, McGraw-Hill, 1935], p. 19.)

[6] See A. Wolf, *A History of Science, Technology, and Philosophy in the 18th Century* 2: p. 452.

[7] Frederick B. Tolles, "Philadelphia's First Scientist: James Logan," *Isis: An International Review Devoted to the History of Science and Its Cultural Influences* 47 (March, 1967): pp. 20–30.

[8] Among these treatises are Johann Hedwig, *Theoria generationis et fructificationes plantarum* (Leipzig, 1798); Auguste P. de Candolle, *Regni vegetabilis systema naturale* (Paris, 1818) 1: p. 71; Ludolph Christian Treviranus, *Die Lehre vom Geschlechte der Pflanzen* in *Bezug auf die neuesten Angriffe erwogen* (Bremen, 1822), pp. 100, 106.

[9] There were several men in early eighteenth-century Massachusetts who experimented with the hybridization of Indian corn, but they did not concern themselves directly with the sexual organs and processes involved in plant reproduction. Paul Dudley (1675–1751) assumed that wind pollination played a role in the sexual generation of Indian corn, but he offered no proof of it. (See Conway Zirkle, *The Beginnings of Plant Hybridization* [Philadelphia, University of Pennsylvania Press, 1935], pp. 104, 130.) Cotton Mather, as early as 1716, described wind pollination as a factor in the sexual generation and hybridization of Indian corn and squashes. His description appeared in a letter to James Petiver, dated September 24, 1716, and he repeated the description in *The Christian Philosopher* (1721). See Conway Zirkle, "Some Forgotten Records of Hybridization and Sex in Plants, 1716–1739," *Jour. Heredity* 23 (October 1932): pp. 433–448.

Another work that escaped Logan's attention was Sebastian Vaillant (1669–1722), *Discours prononcé le juin 1717, a l'ouverture du Jardin royal des plantes, sur la structure des fleurs* (Paris, 1717), translated into Latin (Leyden, 1718, 1728). This is the first writing in France in which proof was established of the sexual

generation of plants. See *Nouvelle Biographie Générale depuis les Temps les plus Reculés jusqu'à nos jours* 45: pp. 834–837. It was Vaillant's work that inspired Linnaeus in his sexual classification of plants.

No evidence has come to light that Logan was aware of the contemporary experiments in Massachusetts.

[10] In a letter to the royal governor of New York, Logan wrote that he had "made some inquiries w[i]th microscopes into the parts of flowers," but that he was unable to determine "the use of their Dust or Male Seed. . . ." (Logan to Burnet, Philadelphia, 18 April, 1728, Letter Books of James Logan, Vol. III, Section C.)

Logan did not see the edition of the *Philosophical Transactions*, in which his letter was printed, until October 1737. He found it very disturbing. As he subsequently wrote to Collinson, his letter had been so edited as to make him "speak nonsense." He complained that additional language had been interpolated into the first sentence of the final paragraph; it was not in his copy of the original letter. He recalled that he entered some marginal alterations in the letter he had sent Collinson, and admitted that he might have forgotten to enter one marginal alteration in his own copy, although he doubted it. What disturbed him was that the alleged interpolation was ambiguous and rendered the meaning of the sentence unclear. Logan urged the publication of an erratum, but no such erratum ever appeared in the *Philosophical Transactions*. It appears that the alleged interpolation had been written by Logan himself in the margin of the original letter, and the entire letter had been printed as received by the Sectretary of the Royal Society, Edmund Halley.

Logan was also disturbed because the letter opened with the salutation "Sir," a word which he certainly would not have used in addressing a fellow Quaker, Collinson. It seemed all the more incongruous in view of the fact that another of Logan's letters, addressed to Sir Hans Sloane and printed in the same edition of the *Philosophical Transactions*, opened with the language, "I shall crave thy Indulgence." Logan suspected that Halley, in editing the letters for publication, had tried to make the Pennsylvania Quaker a fool.

Logan, undaunted by what he believed to be the animosity of Edmund Halley, continued his researches in the role of pollination in the sexual generation of flowers, vegetables, nut-bearing trees, and reeds, as well as maize. Combining his observations and experiments of 1727 and 1728 with his reading in botanical literature, he reached general conclusions which he expressed in the form of an essay in Latin. The essay was published by John Frederick Gronovius in Leyden in 1739 under the title, *Experimenta et meletemata de plantarum generatione*. It was later translated into English by Dr. John Fothergill, an English Quaker physician and friend of Logan and Collinson, and it was published in London in 1747 as *Experiments and Considerations on the Generation of Plants*. Both the English translation by Fothergill and the Latin original are included in this section.

Fothergill, as translator, was very modest. In his advertisement to the 1747 edition he wrote: "The Translator had endeavoured to keep close to his Author's Sense: In point of Expression, he fears, he often falls short of the Original, the Style whereof is nervous, concise, and truly Roman." This appears to have been intellectual modesty in the spirit of eighteenth-century rationalism.

As Professor Frederick B. Tolles has written,

Logan's essay "was a remarkable example of careful and minute observation."[11] By the middle of the eighteenth century it was cited as an authority in much of the botanical literature of the time, in both English America and Europe. Scientific dissertations published in Tübingen and Upsala mentioned it, and Linnaeus, after reading it, thought Logan the greatest botanist of the age.[12]

Logan opened the essay with a description of his experiments with maize. Then he applied his theory of wind pollination to sexual reproduction in the entire vegetable kingdom, drawing from his reading of the experiments and observations of others as well as his own. It will be noted that in describing the sexual parts of plants Logan used terms which are today more commonly employed in zoology. Like Malpighi, he found zoological analogies in the sexual organs and reproductive processes of plants. Such analogies were common in the botanical literature of the time, but for Logan this was not just a matter of terminology. It seemed to him that, if every plant existed in miniature in the semen carried by the air to the female "uterus" where the reproductive process took place, then the same might be true of every living creature in the universe. In the closing sentences of the essay Logan speculated that "the Seeds of all Things," as Anaxagoras had taught, "were in the Air, and descended from thence." Proof of this speculation, Logan suggested, might have as revolutionary an effect on man's conception of the universe as had Copernicus's discoveries of the relationship between the earth and the other heavenly bodies. This speculation was more important to Logan than the contributions he had made to natural history, and explains why he took any interest in botanical experiments at all.

JAMES LOGAN TO DR. WILLIAM LOGAN, PHILADELPHIA, 25 SEPTEMBER 1727, LETTER BOOKS OF JAMES LOGAN, Vol. III, Section A, p. 41, HSP.

Tho' Strictly I Should be more Justifiable in forbearing to write till I hear again from thee, having done it several times within these 12 moneths [months] past without any manner of answer, tho' divers opportunities have offered of late by shipping from thence directly hither yet I cannot omit this from hence w[i]thout a line, not w[i]thstanding thy Silence makes it difficult for me to find anything to say. It may still perhaps be some satisfaction to thee to know we are all in health after a sickly Summer by heats exceeding whatever has been known, in this Country, or as the West Indians amongst us say even

[11] Frederick B. Tolles, *Meeting House and Counting House: The Quaker Merchants of Colonial Philadelphia, 1682–1783* (New York, W. W. Norton, 1963 [The Norton Library]), p. 216.

[12] Frederick B. Tolles, *James Logan and the Culture of Provincial America*, pp. 201–202; Tolles, *Meeting House and Counting House*, p. 217.

amongst their Islands. It began with august and continued a good part of the moneth [month] but is now as cool as is usual for the season.

Last winter I mentioned to thee Woollaston's [sic] Religion of Nature as a Piece I very much admired. In him I took Notice of a hint that Semina primaria Animalium might be received perfect in their kind by the male from without either by the air or food. I have since met with an observation of Bradly's that all plants have their male as well as female seed as he fully proves in the Lilly, 'tis in his new Improvements of Planting and Gardening. from hence I am almost led into an opinion that the first Seeds, the true Essences of all beings exist (perhaps ab origine) perfectly formed in the air & other parts of our Globe, that they are received by the Male in Depositories, fitted for them and there prepared with a proper apparatus to be transmitted into the ffemale Ovaries which in vegetables are what we call their seeds. This gives me a field of Contemplation that wonderfully entertains me, and leads me to look upon the Creation in a manner very Different from what ever had entred my thoughts before. I wish I could have thy thoughts on it, and particularly that the Botanists would search for this Semen Masculum through all the vegetables that are known. If this be the Case the Creation is vastly different from what has hitherto been imagined.

. . . .

James Logan to Thomas Goldney, Philadelphia, 20 November 1727, Letter Books of James Logan, Vol. III, Section A, p. 43, HSP.

I answered thine by thy Son in a few humorous Lines by himself at his return, w[hi]ch I hope for his sake will come safe to hand, before this & tho' trifling, will not prove altogether unacceptable.

My intention in this is to putt thee, as a Botanist, on giving thy Self a piece of Entertainm[en]t next Spring & Summer in thy Garden, w[hi]ch I am persuaded will yield thee, as agreeable a Speculation, as most things thou hast mett w[i]th in nature either there or elsewhere.

I suppose it cannot be unknown to thee that of late years a Discovery has been made that there is a male as well as female seed in all vegetables, that the male is the true seed, & th[a]t the noted seed vessels are ovaries every grain of w[hi]ch being impregnated becomes what is commonly known by the name of seed.

The 2d Chaptr. of th[a]t Book of Rd. Bradly's called new Improvem[en]ts of Planting & Gardening, is wholly on this Subject, & he fully demonstrates the Truth of it by the Experim[en]ts he there relates. His Instances are but very few, tho I think there is no Subject in Botany that better deserves to be inquired into, & pursued through the whole families of Plants, nor any thing that can be more curiously entertaining.

In hopes th[a]t thou hast, or may meet w[i]th the Book, I shall not here repeat any thing he has said further than to observe, that in all Plants, at the same thime, th[a]t the flower breaks out there is found, in some part of the same Plant (except where one Plant is wholly male & others female, as is said of the Palm tree some Lettices &c) a loose Dust that flies off from the place where it appears to be bred & falling on the Pistil or some thing equivalent to it that gives a passage to the Ovary, impregnates the tender germens there, invigorated by which they afterwards advance to maturity.

This Dust, as it appears is the true male seed, & is most commonly within the blossom or flower, but frequently w[i]thout, as in the Instance I am going to give thee, for the sake of which I principally write this letter.

Our Indian Corn or Maize is a very singular Plant, but now common in the Gardens of the Curious w[i]th you. When the Stalk of this is grown to its full Stature, w[hi]ch in a good Ground & well dress'd is about 8 or 9 foot high, there breaks out of the top a kind of Tassel, of divers thick Strings, of w[hi]ch till I mett w[i]th this new thought, I never could imagine the use. After these are fully blown, the heads break out below in the joynts between the Leaves, & the Stalk, w[i]th a kind of flower at the top made up of a great number of fine threads red or yellow curled & frizled, & just about the same time these appear, the tassel or tufted top sheds a Dust from it, like that from the catkins of Hazels & divers other Trees.

But the most remarkable part is, that on opening one of these young heads before the fine threads I have mentioned wither, we find them all lie perfectly straight, Like a Stretched Skin of fine white Silk, till they came as low, as the corn head, & th[e]n parting themselves into as many divisions as there are rows of the Corn, w[hi]ch are most commonly eight, but in rank Land sometimes 10, or 12, & passing along between these Rows their roots enter at the roots of the Corn, one of them to each grain in exact order from top to bottom, & just so many Grains as there are in the head so many strings there are in the flower, that is from 300 to 5 or 600, & thus every single thread is a distinct pistil to convey the male seed into the ovum w[hi]ch when grown becomes in this as in all other the like cases, what we call (as I have said) the seed it self.

The same perhaps are the awns of barley wheat Rye oats &c but then I have never yet sufficiently observed and in general I must say, that I think there can scarce be found a finer Diversion in nature than to discover and view the various appartus [apparatus] for this kind of Generation through all the different Sorts of Vegetables, th[a]t the spring season presents especially in well furnished gardens &c.

But to discover how this male seed is formed whose every smallest Grain scarce distinguishable to the bare eye, contains the Rudiments or Stamina of a perfect plant, in all its parts, will require a penetration far exceeding mine, tis obvious that it appears to be a kind of adventitious Dust scarce belonging to the plant, as if lodged there like Dew from the Air.

This would lead into a Philosophical Inquiry whether tis possible to imagine how, either a vegetable or animal Body w[hi]ch are now well known to be nothing else than a contexture of fibrils & canals w[i]th liquids in them more or less, & th[a]t their Life, is nothing but the motion of these Liquids[.] I say how from this Life or motion, the stamina of a fetus of any kind can be produced organized & divided into all its distinct parts is a point in Philosophy, th[a]t has ever been beyond the reach of my Conception, but admitting such a formation possible in the Plant, how such vast numbers of those organiz'd bodies viz many millions on some Trees (for the Dust when shaken will make a little Cloud) all sticking to the outermost parts of the Plant, as if they scarce belong'd to it, adds greatly to the difficulty how to account for it, as I noted before, & might lead one to believe, th[a]t the true Seeds of all productions or the beings themselves exist somewhere else before they are cloathed with a visible corporeity. But this is a Speculation, th[a]t perhpas is never to be comprehended, by the limitted Intellects of Mankind. It may serve however, to teach us to be modest, & admire the Power by w[hi]ch the whole creation is conducted, & still yields a large scope to entertain our minds, w[i]th what is really visible & intelligible in the wonderful Process.

If thou wouldst try any Experim[en]ts on this Indian Corn, w[hi]ch are to be made I believe on that Plant as advantageously as on any other Whatsoever, Be pleased to mix some of the Garden mold w[i]th a little fer[tilizer] [page torn] low house or stable, for such Dung best agrees w[i]th it, making [page torn] hill or two in several distant parts of thy Inclosure, open these little [hills?] in April, when the air begins to warm, & scatter in them 3 or 4 Grains at a few inches distance, then cover them up about 3 or 4 inches deep w[i]th the Earth, keeping it clear of all weeds, during all the time of its Growth[.] when about a foot high loosen the earth w[i]th a Howe, & again when it begins to push out its top or tassel and to make thy Trials of the Generation I have mentioned, In one hill thou may nip off all these Tassels as soon as they appear, & then if I mistake not, thou wilt find these Stalks, tho' they push out their young heads below yet they will not fill or bring any ripe Corn. In other Hills letting the Tassels Stand, thou may pinch off some of the flowers of threads, just as they are appearing on some heads, taking them intirely off, on others taking only a part, & leaving the rest. And then in those, that are quite pinched off thou wilt I believe have no Corn, or in the others perhaps as many Grains, as thou hast left threads standing & unhurt. w[hi]ch if it succeeds as I have said will give thee a full Demonstration of what I have hinted.

On the late Prospect we have had of Peace in Europe, I had thought to bring over my family to Engl[an]d next summer, but that appearing more Doubtful at present, I am in Suspence also as to my voyage. If I come & thou makes these Trials, thou may then acquaint me w[i]th the Issue, or if I come not, then if thou please by a Lett[e]r, w[hi]ch upon any Subject will always be very acceptable from thee. . . .

James Logan to Alexander Arscott, Philadelphia, 29 November 1727, Letter Books of James Logan, Vol. III, Section A, p. 45, HSP.

About a month since I took the freedom in a Letter via Lond[on] to impose a double Charge on thee, w[hi]ch I know not well how to justify otherwise than by alledging th[a]t the manifest proofs of thy friendship to me at all times since our acquaintance, but especially, when I was last at Bristol Seem'd to point thee out to me as the person I could be most free with in matters th[a]t my Brother could not properly be employ'd in. . . . I requested G Goldney,[13] whom I saw after my Lett[e]r was sent on board, to desire thee to take no notice, till thou heard further from me, for I had just then rec'd Lett[e]rs from Engl[an]d, w[hi]ch cast such difficulties in my way, as p[er]haps might render the Proposal of a Voyage impracticable to me, & if safe arrived, as I hope he will be w[i]th you long before this, I doubt not but he has answer'd my Request.

I have indeed of late flatter'd my self w[i]th the hopes of Spending Some part of the Remainder of my time, whatever th[a]t may be, w[i]th my fr[ien]ds in Engl[an]d, but the Prospect decreases, & it now seems more probable that I shall never in this Life change this Climate for another.

. . . .

I have by this Ship wrote a long Lett[e]r to T Goldney, of a wonderful late Discovery in nature, w[hi]ch tis probable thou may see & if it be new to thee, thou may find matter in it for very deep Speculation, beyond whatever in a common way has faln under humane consideration.

That excellent author Wollaston in his Religion of Nature seems to have had the Notion when he hints it as a probability th[a]t all male animals take in the animalcula of the Semen from the air or food &c.

. . . .

James Logan to Col. Burnet, 10 March 1727/8[?], Letter Books of James Logan, Vol. III, Section C, p. 191, HSP.

. . . .

I am troubled . . . to find Sir Isaac's [Newton's] opinion was that the animal is in the ovum, for if the

---

13 G. Goldney was Gabriel Goldney, the son of Thomas Goldney.

new discovery of the male seed in vegetables universally holds, I cannot believe but it will amount almost to a Demonstration that the cause is otherwise, and why a Dust from the summets or extream parts of Plants Trees &c should be wanted to fecundate the ovum lodged carefully in a well guarded Ovary or Uterus may p[er]haps be found inexplicable, otherwise than by allowing that the true semen it self, the formed Plant is thereby convey'd[.] towards the knowledge of w[hi]ch proper experiments on our Indian Corn, as I hinted before may afford matter both for Speculation & Instruction not only by lopping off the Top blossoms from some planted remote from all others lest the wind should convey the Dust but more particularly in pinching off a part of the threads w[hi]ch make the flower at the top of the Corn head at their very first appearance & then observing afterwards whether so many of the grains in that Head doe not miscarry & fail of filling. But what I principally consider in the Case is what may be suggested from that hint in Wollaston for w[hi]ch I referrd to the page.

. . . .

JAMES LOGAN TO DR. WILLIAM LOGAN, PHILADELPHIA, APRIL 8, 1728 (postscript), LETTER BOOKS OF JAMES LOGAN, VOL. III, SECTION C, p. 200, HSP.

I could now wish, Since I have declined my intended voyage to England, that I had not wrote the long Lett[e]r I sent Tho: Goldney last Fall about the Propagation of Vegetables which I suppose thou may have seen. At that time, I expected to be at Sea or in Britain during this ensuing Summer & was willing to putt him, as a Person not unskill'd in Botany, on making those entertaining Experim[en]ts especially on our Indian Corn, w[hi]ch by reason of its multitude of fine threads, each serving for a distinct Pistillum to every several Grain, and on divers other Considerations, particularly the distance of the top blossom from the heads of Corn themselves, I take to be the fittest vegetable for tryals of that kind, that I know. I am sensible they may be made on all other sorts of headed Grain, the awns of w[hi]ch I take also to be so many pistillums, but none of them afford opportunities equal to the other, yet it may be worth while to observe, when their blossoms first appear, just as 'tis forming, how these awns are form'd at first before they shoot out, to a length above the head.

I have tried divers flowers this Spring already, & particularly the blossoms of Some trees & every where find an apparatus agreeing w[i]th the hypothesis, but in the apices of those that produce only one seed or fruit, for one blossom, as the Cherry Peach &c. I can discover (I must own) nothing to answer the Semen masculum lodged on the apices, besides find transparent globules, adhering to them, w[hi]ch are also very few in number, while in others that produce a knott, or case of seeds, under each flower, as the

tulip of w[hi]ch we have one early sort here already in flower, the apices are cover'd plentifully with a dust, each particle of which is form'd somewhat like a Grain of wheat, but more acuminated, at both ends, like a plum tree leaf & afford a very entertaining Spectacle.

I have seen an advertise[ment] of a Book, printed for, or sold by Innys published by one Patrick Blair, called Botanick Essays, concerning the structure of flowers, the fructification of Plants, w[i]th their Generation, their sexes & manner of impregnating the Seed &c 8ᵛᵒ 1720, w[hi]ch if it answer the title, I should much desire to see, & therefore design to Send for it by the first opportunity. Pray Look into it, & say something to me on that head in which thou wilt oblige me.

SOME EXPERIMENTS CONCERNING THE IMPREGNATION OF THE SEEDS OF PLANTS, BY JAMES LOGAN, ESQ; COMMUNICATED IN A LETTER FROM HIM TO MR. PETER COLLINSON, F. R. S., PHILADELPHIA, NOV. 20, 1735. *Phil. Trans.*, **38**, 435: pp. 441–450.

Sir

As the Notion of a Male Seed, or the Farina Fecundans in Vegetables is now very common, I shall not trouble you with any Observations concerning it, but such as may have some Tendency to what I have to mention—— And, first, I find from Miller's Dictionary,[14] that M. Geoffroy,[15] a Name I think of Repute amongst Naturalists, from the Experiments he made on Mayze, was of Opinion, that Seeds may grow up to their full Size, and appear perfect to the Eye, without being impregnated by the Farina, which possibly, for ought I know, may in some Cases be true; for there is no End of Varieties in Nature:—— But in the Subject he has mention'd I have Reason to believe it's otherwise,

---

[14] This appears to be a reference to Philip Miller (1691–1771), *The Gardeners Dictionary: containing methods of cultivating and improving the kitchen, fruit and flower-garden. . .* (London, Printed by the author, 1711). A second edition was published in London in 1733. An abridgment from the folio edition was published in London in 1735. Logan's reference is to either one of the folio editions. This was the most authoritative garden encyclopedia in the eighteenth century.

[15] Geoffroy, a seventeenth-century botanist, presented a memoir to the Academy in Paris on plant anatomy and the stamens as male organs. This was after Nehemiah Grew presented his paper on plant anatomy to the Royal Society of London in 1676. (Sir William Dampier, *A History of Science and Its Relations with Philosophy and Religion* [Cambridge University Press, 1958], p. 167. Etienne-François Geoffroy (1672–1731) was the son of an apothecary who had originally intended him for a pharmaceutical career. Geoffroy passed the pharmaceutical examination, but he continued his medical studies and obtained a doctorate in 1704. In 1709 he was appointed to the chair of medicine and pharmacy at the Collège de France. He published a number of books on medicine, materia medica, and chemistry between 1704 and 1725. Although he was chiefly interested in botany and chemistry, he had an enthusiastic interest in all the basic sciences. See *Nouvelle Biographie Générale depuis les Temps les plus Reculés jusqu'à nos jours* **20:** pp. 30–33.

and that he applied not all the Care that was requisite in the Management.

When I first met with the Notion of this Male Seed, it was in the Winter Time, when I could do no more than think of it; but in the Spring I resolved to make some Experiments on the Mayze, or Indian Corn. In each Corner of my Garden, which is forty Foot in Breadth, and near eighty in length, I planted a Hill of that Corn, and watching the Plants when they grew up to a proper Height, and were pushing out both the Tassels above, and Ears below; from one of those Hills, I cut off the whole Tassels, on others I carefully open'd the Ends of the Ears, and from some of them I cut or pinch'd off all the silken Filaments; from others I took about half, from others one fourth and three fourths, &c. with some Variety, noting the Heads, and the Quantity taken from each: Other Heads again I tied up at their Ends, just before the Silk was putting out, with fine Muslin, but the Fuzziest or most Nappy I could find, to prevent the Passage of the Farina; but that would obstruct neither Sun, Air or Rain. I fastened it also so very loosely, as not to give the least Check to Vegetation.

The Consequence of all which was this, that of the five or six Ears on the first Hill, from which I had taken all the Tassels, from whence proceeds the Farina, there was only one that had so much as a single Grain in it, and that in about four hundred and eighty Cells, had but about twenty or twenty one Grains, the Heads, or Ears, as they stood on the Plant, look'd as well to the Eye as any other; they were of their proper Length, the Cores of their full Size, but to the Touch, for want of the Grain, they felt light and yielding. On the Core, when divested of the Leaves that cover it, the Beds of Seed were in their Ranges, with only a dry Skin on each.

In the Ears of the other Hills, from which I had taken all the Silk, and in those that I had cover'd with Muslin, there was not so much as one mature grown Grain, nor other than as I have mentioned in the first; But in all the others, in which I had left Part, and taken Part of the Silk, there was in each the exact Proportion of full Grains, according to the Quantity or Number of the Filaments I had left on them. And for the few Grains I found on one Head in the first Hill, I immediately accounted thus: That Head, or Ear, was very large, and stood prominent from the Plant, pointing with its Silk Westward directly towards the next Hill of Indian Corn; and the Farina, I know, when very ripe, on shaking the Stalk, will fly off in the finest Dust, somewhat like Smoak. I therefore, with good Reason, judged that a Westerly Wind had wasted some few of these Particles from the other Hill, which had light on the Stiles of this Ear, in a Situation perfectly well fitted to receive them, which none of the other Ears, on the same Hill, had. And indeed I admire that there were not more

of the same Ear than I found impregnated in the same manner.

As I was very exact in this Experiment, and curious enough in my Observations, and this, as I have related it, is truly Fact, I think it may reasonably be allowed, that notwithstanding what Mr. Geoffroy may have deliver'd of his Trials on the same Plant, I am positive, by my Experiment on those heads, that the Silk was taken quite away, and those that were cover'd with Muslin, none of the Grains will grow up to their Size, when prevented of receiving the Farina to impregnate them, but appear, when the Ears of Corn are disclosed, with all the Beds of the Seeds, or Grains, in their Ranges, with only a dry Skin on each, about the same Size as when the little tender Ears appear fill'd with milky Juice before it puts out its Silk. But the few Grains that were grown on the single Ear, were as full and as fair as any I had seen, the Places of all the rest had only dry empty Pellicles, as I have described them; and I much question whether the same does not hold generally in whole Course of Vegetation, though, agreeable to what I first hinted, it may not be safe to pronounce absolutely upon it, without a great Variety of Experiments on different Subjects. But I believe there are few Plants that will afford so fine an Opportunity of observing on them as the Mayze, or our Indian Corn; because its Stiles may be taken off or left on the Ear, in any Proportion, and the Grains be afterwards number'd in the Manner I have mentioned.

LOGAN TO JOHN BARTRAM, *ca.* 1736, IN ARMISTEAD, *Logan*, pp. 157–58.

Last year, Peter Collinson sent me some tables, which I never examined till since I last saw thee. They are six very large sheets, in which the author (Luinious)[16] digests all the productions of nature in classes. Two of them he bestows on the inanimate, as stones, minerals, earths; two more on vegetables, and the other two on animals. His method in the vegetables is altogether new, for he takes all his distinctions from the *stamina* and the *styles*. He ranges them under those of one stamen, 2, 3, 4, 5, 6, 7, 8, 9, 10, 12, 20 and then of many stamens. He further distinguishes by the styles, and has many heads under which he reduces all the known plants.

This performance is very curious, and at this time worth thy notice.

JAMES LOGAN TO PETER COLLINSON, STENTON, OCTOBER 31, 1737. ROYAL SOCIETY OF LONDON.

When I wrote my last of the 10th Inst[ant] via Bristol I design'd not to give thee any further trouble of the kind before Capt[ain] Wright should Sail

---

[16] This is undoubtedly Karl von Linné (Linnaeus). The tables were the first edition of Linnaeus's *Systema Naturae*, 14 pages folio, of 1735.

which he intends this ffall.   But as it is not impossible that he may be stopt by the Ice, as it has often happen'd to others, having yesterday rec[ei]v[e]d thine of the 12th (I suppose of Aug[us]t tho the Month is omitted) p[er] Capt[ain] Linsey with two Philos-[ophical] Transactions, I shall here acknowledge them and briefly Speak to the rest of thy Lett[er].

Having compared what is printed of mine in N. 440 with the Copy I have of my Lett[er] to thee from whence it is taken, I am Surprized at a variation which makes me Speak nonsense.   My Copy of the Paragraph beginning at the bottom of pa 194 of the Print runs in these words.   "As I was very exact in this Experim[en]t and curious enough in my observations, and this, as I have related it, is truly fact, I think it may reasonably be allow'd, that notwithstanding what M. Geoffry may have del[i]v[ere]d of his trials on the same plant, yet none of its grains will grow up to their Size, when prevented of receiving the farina to impregnate them.   The few grains that were grown in this Single ear were as full and fair as any I had seen, the places of all the rest had only dry empty pellicles &c as in the Print."[17]   I am therefore much at a loss to think how the interpolation that appears in the Print must have happen'd.   I have formerly told thee that what I sent thee was my first draught.   When finished it was fairly copied by my Son, but I remember I made some alterations in the original both before and after it was copied, which last now appear in my own hand in my Copy, and if I made so material an one as that in the print I cannot enough admire how I could neglect to make the Same in the Copy I was to keep.   But what affects me is that as it now stands in the print it is nonsense, for is it reasonably to be allowed that I am positive by my experim[en]ts on those Heads?   Or am I positive that the Silk was taken quite away? I am apprehensive indeed that the mistake is owing to something I may have added in the margin of my Lett[e]r w[hi]ch I forgot to add also to the Copy, but Surely I Could not myself make such nonsense of it, or if I did, it was certainly incumbent on those who excerpted out of & altered what I wrote as they judged most proper, to have made good Sense of what they published, in which I am really concerned to find they have Strangely failed.   Had it been published thus it might have read tolerably viz (at the last line of the page) "the Same Plant (as I am positive, by my experiments on those Heads that the Silk was taken quite away from, and those that were covered with Muslin;) none of the grains will grow &c pointing it as I have done, or rather "on those Heads from which the Silk was taken quite away, & those &c."   And if the Secretary who published that Extract would make and note such an Alteration in the next Errata he would oblige me, tho it is no more than a piece of

[17] Compare with Royal Society, *Phil. Trans.* **39**, Num. 440: pp. 194–195.

common Justice.   But pray how came this Extract from my Lett[e]r to thee to be made to begin with Sir which I am sure was not there, while in the other from my Letter to Sir Hans Sloane I begin with "I shall crave *thy* indulgence?"   This is making me inconsistent with my Self, & exposes me.

.   .   .   .

[Memorandum added by Peter Collinson: Mr. Logan's Letter, which is entered in Letter Book of R. S. vol. 21 pag. 241 is enter'd as the author first sent it.   I agree with the Corrections he here makes of its publication in Phil. Trans.]

De Plantarum Generatione Experimenta et Meletemata Leyden, 1739.

Cum de generatione tum stirpium, tum animalium, olim mihi variae suborirentur dubitationes, quamprimum de farine foecundante, seu semine masculino inaudiveram, spes mihi haud mediocris affulsit, lucis aliquid rei adeo obscurae inde affundi posse.   Et cum jampridem vix citra stuporem crescendi modum in planta *Americana*, quae frumentum *Indicum*, seu *Maiz* nuncupatur, observâssem, hanc omnium quas noveram, vel forte quas educat natura, experimentis capiendo maxime idoneam censebam.

Ad proceritatem enim sex, octo, quandoque decem pendum evehitur: in culmi vertice stamineum fert cirrum, seu (uti vocat *Malpighius*) muscarium apicibus instructum, e quibus evolvitur farina; subtus ex caulis geniculis spicas protrudit, sex, octo, decem, vel etiam duodecim pollices longas.   Spica ex scapo interraneo bene solido, et pollicem circiter crasso constat, cui adnascuntur pulcherrimi granorum ordines, plerumque octo, saepe autem decem, rarius vero duodecim; sed etiam sedecim vidi.   In quovis ordine sunt semines grana plus minus quadraginta, quae in formationis primordiis, cum adhuc tenella est spica, ova jure dici possent: cuique autem ovo adnascitur filum album, laeve, tenue, et, praeterquam quod sit cavum, fili sericei instar: haec singula fila inter ordines seriatim a prima ad extremitatem ulteriorem perreptant, ubi ex discludentibus se foliis, quae spicam totam muniunt in fasciculo seu schoeno, se aëri sistunt prominentia, colore saepius, in hac parte prominente, albicantia, quandoque vero pro varia plantae specie, flaventia, rubentia, aut purpurascentia; atque haec fila, ut suspicabar, veros ovorum stylos esse postmodum deprehendi.

In hac planta igitur experimenta capturus, in hortuli mei urbani quadraginta pedes lati, et octoginta circiter longi singulis angulis, colliculis pro sationis more aggestis, mense *Aprili* inclinante, seminis grana quaterna aut quina commisi.   Inchoante autem *Augusto*, cum planta ad justam magnitudinem jam increvissent, et cirri in vertice, spicae ad caulem se plene extulissent; ex uno colliculo totos hos cirros penitus demessui; in alio autem intactis cirris, ab aliis spicis lenissime disclusis eas obtegentibus foliis, et denuo conclusis, totum filorum, seu stylorum fasciculum abscidi; ex aliis tantum unam, ex aliis binas, ex aliis ternas partes, quartas, reliquis integris relictis: aliam vero spicam, priusquam se luci sisterit fasciculus linteo *Indico* seu *Sinensi*, tenui et lanuginoso (*Muslin*) levissime obvolvi, ita vero laxe, ut vegetationi ne minimae esset fraudi, quin propter lintei tenuitatem solis, aëris, et imbrium beneficio spica frueretur, nullus vero propter lanuginem pateret farinae aditus. Quarti colliculi plantas integras et intactas, ut et aliarum quamque in eo quem dixi statu ad maturationis temptestatem permisi.

Adventate igitur *Octobri* mense, cum in operationum harum eventum inquirendi afforet tempus, haec mihi comperta sunt. In primo colliculo, ubi omnes decollaveram cirros, licet spicae speciem satis commodam conspectui exhibirent, involucris suis utpote (ut semper) obtectae, tactui tamen graciles et leves, ne vel unicum perfectum granum reperi, praeterquam in una spica grandiore, quae e caule suo protensius eminens, adversus colliculum proximum (ex qua parte apud nos vehementiores spirare solent venti) prospectabat; in ista enim unica spica viginti praeterpropter ad justam molem et perfectionem ematuruerant, granulorum farinae ope, ut conjiciebam vento isthuc aliunde delatorum. In iis quibus evulseram stylos, totidem ad unum reperta sunt semina, quot reliqueram intactos stylos; in iis quas linteo obvinxeram, ne vel unicum; in ovis irritis seu cassis, nihil praeter pelliculam subaridam conspiciebatur.

Ex his experimentis a me summa cum accuratione institutis et patratis, ut et ex aliorum plurimis, farinam hanc ex apicibus evolutam verum semen esse masculinam, et faecundationi uteri et seminum omnino necessariam liquidissime constat; quod tamen omnia secula usque ad nostrum letebat. Dignus jure merito est igitur hujus naturae arcani inventor, cujus perpetua celebretur memoria. Is autem fuisse videtur *Thomas Millington*, eques auratus *Anglus*, professor suo tempore *Savilianus*, ante vel circiter annum 1676. ita enim retulit *Graevius* (*Nehemiah Grew*, M. D.) in lectione coram *Societate Regia*, nono *Novembris* istius anni habita. (Vide *Graevii*, opera, p. 161, et 171.) *Malpighius* quidem, quantum novi, nusquam illius usus meminit; ipse autem *Graevius* ad foecundationem necessariam esse farinam, non autem uterum ingredi suspicatus est: at viginti aut plus eo post annos, *Samuel Morlandus*, itidem *Anglus*, in ipsum uterum per styli canaliculos delabi asseruit. (Vide *Acta Philos, Angl.* num. 287.) Ego quidem semel in medio unius ex supra memoratis stylis frumenti *Indici* granulum conspexi: nec dubitandum reor, quin exquisitiore adhibita diligentia, in iis delabentia facile deprehendantur. Hactenus autem dicta jam satis sunt vulgata, et harum rerum curioso nemini non nota; quae vero sequuntur observata eorum meditatione qui naturae indagandae incumbunt, non indigna fortasse

comperientur. Post haec capta experimenta, mihi neque per valetudinem, nec per negotia, his disquisitionibus vacare fere licuit: quoad vero licuit ista praecipue notanda mihi videbantur.

Non tantum in hac planta, in arboribus nuciferis, arundinibus, cucurbitus cum tota earum familia, scilicet, pepone, melone, cucumere, &c. in quibus omnibus dissitas occupant sedes hae partes masculae et foeminae, quae generationi inserviunt, sed in eorum etiam florum plerisque, qui a quibusdam hermaphroditae nuncupantur, quippe quod intra idem petalorum septum utrasque partes complectuntur, eo locantur situ apices, ut postquam elaborata fuerit farina, vix, et saepius ne vix quidem, os uteri seu styli summitatem possint attingere: sed in iis etiam pariter ac in longius dissitis, necesse habetur, ut farina prius ex sede sua in apice, divulsa penitus et eliminata, libero innatet aëri undique illam ambiente, et casum quasi fortuitum subeat, num ad os uteri seu styli unquam perveniat (quo tamen ut perveniat, atque in eo fungatur officio suo, ut fructum fera planta, omnino exigitur) aut vento et imbre discussa diffletur, et prorsus dissipetur.

Huic affinis etiam videbatur insignis ea apparatus discrepantia, quem ex altera parte in compangendo et tutando utero cum ovis suis, et ex altera in formandis staminibus, apicibus, et farina adhibitum a natura cernimus. Uterus seu ovarium ex ramusculi seu cauliculi meditullio educitur, et pro plantae indole validissime undique munitur, cum ipsa tamen ova sua usquedum a semine masculino foecundentur, ex sola pellicula liquore homogeneo repleta constare videantur: at vero stamina huic semini foecundanti ferendo distincta, vel tantum ex deciduis florum petalis, tanquam ex cute pili, vel una cum iis ab eadem origine enascuntur: quod videre est in iis etiam floribus, quorum styli, ut in malvis, &c. theca aut vagina staminea muniri conspiciuntur; quae duo, scilicet stylus et vagina, licet rem unicam conficere videantur, omnino tamen dispar est partium origo: stylus enim, ut in omnibus, e medio exsurgit ovario, vagina vero petalis, utpote ex iisdem fibris oriundis adhaeret. Apices autem ex pelliculis pulpa quadam, ut videtur, parenchyma et homogenea refertis constant, quae tamen pulpa, dum adhuc mollis et uvida, in granula sua dividi comperitur. Horum apicum paulatim siccescentium tunicis, quae sunt tenuissimae, disruptis evolvitur farina, quae ita tunicis adhaeret, ut primo vix ulla concussione a sede sua dimoveatur: at cum apices ex tenui isto filo penduli in aëre motitentur, sensim sponte defluit, et floris variis partibus, praecipue autem styli summitati, quae liquore viscoso aut lento instruitur (terebinthinam vocat *Malpighius*) tanquam pulvisculus temere inspersus adhaeret. Sin autem per aliquot dies serenos nulla spiret aura quae apices moveat, ex caulis frumenti *Indici* concussione, in altum se feret pulvis, nebulae aut fumi instar. Hoc etiam in hac planta notari meretur, quod quo die disrumpuntur apices, et pendulos se produnt in aëre, eodem die cernuntur congeriei seu fasciculi stylorum extremitates, ex spicae integumentis se in aëra protrudentes, quod in aliis etiam observandi ansam merito praebere possit.

Farinae hujus granula in eadem plantae specie sunt omnino uniformia; at in diversis diversam obtinent figuram. In plurimis sunt teretia seu globosa, in multis oblonga tritici instar, in quibusdam, veluti caltha, malvis, et aliis, rotam dentatam seu globulum aculeis obsitum repraesentant. Angularia nunquam vidi, qualia tamen inter alias formas exhibet *Graevius* in viola tricolore. (*Anatome*, Tab. 58.) Sed plerumque sunt laevia; et nitentia in frumento *Indico* compressa videntur.

Sed in ipsa seminis generatione seu incrementi sui processu, uti eum enarrarunt et depinxerunt celeberrimi illi plantarum anatomici, quod etiam partim ex autopsia est mihi comprobatum, sunt quae in admirationem plane rapiant, et quae meditatione penitiore jure digna videantur.

Primo enim cum defluente omni floris satellitio, et emarcescentibus stylis, intumescere coeperit ovum, in ejus summo prodit se vesicula, quam amnion appellat *Malpighius*, umbilicali instructa per chorion in adversam ovi partem producto. Augescente vero cum ovo amnio, post aliquot dies in hujus vertice alterum quiddam minutulum conspicitur, quod itidem indies increscit, usquedum totum amnion, ut hoc totum chorion, ovumque totum compleat; his duobus (amnion et chorio) in putamina aut seminis tunicas cedentibus: et hoc jam ad justam suam molem auctum et maturatum pro vero semine habetur, quod intra putamen et tunicas plerumque in binos, rarius in plures lobls dispescitur, quandoque vero in nullos, ut in frumentis, in quibus nulla est hujusmodi divisio.

Sed in hoc processu, prae caeteris hoc notandum occurrit, nempe quod dum increscit semen, in apice ejus, ut in arboribus nuciferis, pomiferis, et aliis; in quibusdam vero ad latus, aliud se prodit minutius quiddam, quod primo tanquam granulum adventitium, vix ullibi cum reliqua seminis mole cohaerens deprehenditur: sensim autem tot projicit venulas quot sunt in semine lobi, per quas se iliis tenuiter annectit, et hae per lobos dispersae, ut arteriae in foliis diradiantur.

In seminibus quorum apici hoc insistit, res plane distincta et a semine diversa esse videtur: constat autem duabus partibus altera, quae detractis exterioribus integumentis conspectior prominet, a *Graevio* radicula dicitur, quod in radicem plantae abeat; altera vero, quae inter seminis lobos delitescit, ab eodem pluma nuncupatur; nam quod cauliculum cum foliis in se contineat, plumosa videtur: utraeque autem partes quodam internodio conjunguntur.

Haec pluma in seminibus grandioribus, quorum lobi e terra cauli adhaerentes sursum feruntur, et in folia evadunt seminalia, divulsis lobis, conspicuam (et in phaseolis amoeno quidem spectaculo) se praebet.

In favis, et pisis, et aliis hujusmodi quorum non exsurgunt lobi, crassior et firmior videtur hujus cum lobis nexus; at detracta secunda seminis tunica, quae hujus est prima seu exterior, eodem modo se habere cernetur quo se habet in nuciferis aliisque jam memoratis, extraneum nempe aliquid et institutium a toto reliquo semine alienum videbitur: et hoc verum esse plantae seminulum, seu potius veram plantulam, ad vegetationem in alterum annum jam conformatam et pararam constat. Reliqua vero tota seminis moles, tum animalium victui, tum alimento plantulae idoneo subministrando, donec firmiores in terra egerit radices, solummodo inservit.

Ex hoc seminulo solo totam educi plantam, nisi quod caulis tunica exterior ex interiore seminis tunica quandoque constituitur, certissime liquet. Namque licet in frumentis quibusdam, ut hordeo, avena, frumento *Indico*, radicula ex altera parte seminis in radicem abeat, ex adversa vero parte emergat caulis; hic caulis veruntamen haud secus ac in aliis, ab eadem prodit radicula; ab ea enim per venam seu canaliculum subscutaneum perrepit, donec ex altera parte se exserat, nec in progressu suo, cum reliquo semine quicquam videtur habere commune.

Ita igitur investigato hoc generationis processu, reliquum est, ut quid inde verisimillimum colligendum sit aut astruendum, dispiciamus. Primo farinam utero aut ovis foecundandis omnino necessariam vidimus: huic item farinae producendae tenuem admodum a natura adhiberi apparatum: maturatam vero ventis et aëri, priusquam os uteri seu stylum possit attingere, plerumque permitti: cui processui, a perpetua naturae in aliis lege, toto coelo discrepanti, certam aliquam subesse rationem minime dubitandum est. Naturam enim brevissimum et certissimum operandi modum affectare in omnibus constat: itaque in hoc negotio farinam utero ingerendi, ni obstaret majoris momenti aliquid, cui aliter providendum erat, expeditiora viae compendia instituram fuisse existimandum est. In ovis porro, antequam a farina impraegnentur, nihil praeter liquidum quoddam (ut vocat *Malpighius*) colliquamentum reperiri cernimus: in amnio novam moleculam se exserere, quae totam constituit seminis molem: in hoc semine denique tanquam omnino extraneum et insititium quiddam, verum seminulum, seu veram plantulam comparere.

Jam vero si cum totum hunc processum, tum ipsius plantae conformationem, ex quot nempe et quam variis ea constet partibus, radice scilicet, caule, foliis, canalibus, colis, cum infinitis aliis rite perpendamus, unde etiam vel quibus modis et meatibus in hoc granulum convehi possint tot diversarum partium stamina, et quo intercedente artificio, in stupendum illum ordinem quem cernimus et ornatum digerantur et compaginentur, sedulo reputemus; vix est ut ex isto simplici processu hanc artificiosissimam conformationem et compaginationem plane mirabilem quovis pacto deduci posse existimemus: non enim tantum humanum (ut pleraque) captum superare, sed rationi etiam absonum meriot possit videri.

Itaque cum ab omni aevo hoc generationis opificium indissolubilibus premi difficultatibus compertum sit, omniaque illius explicandi molimina hactenus incassum cesserint, quidni ipsam naturam in isto processu, in eo praesertim quod seminulum a reliquo semine discretum ita obvium oculis nostris sistat, rem totam quasi digito palam monstrare censeamus? Farinam scilicet ideo aëri liberam dimitti, quo ipsissimum verum seminulum, seu plantulam extrinsecus in staminibus suis minutissimis, et quidem ob exiguitatem aciem nostram fugientibus omnino tamen perfectis, jam prius formatam et paratam in se recipiat, et hujus virtute praegnans reddita vi indita attractrice, ut mares et foeminas in animalbus, se invicem etiam e longinquo appetere cernimus, primum in stylum, in eoque per canaliculos sibi proprios usque ad ova delabatur ex hac autem farina plantae succis enutrita, ad usus, quorum supra meminimus, seminis confici molem in eo dein latitantem plantulam terrena jam materia ex farina indutam, se exserere, et tandem alimentis e terra idoneis adaucta palam se efferre sub auras.

Nec in hac stirpium generatione sola fortasse haerendum est. Sed etiam ad omnium quae in orbe nostro vita donantur universitatem easdem cogitationes proferre aequum et rationi consonum videri possit. "Omne scilicet quod in hac tellure vitale est (tum stirpes tum animalia) prius staminibus suis integre efformatum existere, atque ex aëre seu aethere deduci; et hic materia terrena convestitum augescere et dilatari: neque aliud esse generationem quam seminis hujus mari seu masculo immissi, et in eo indolem terrestrem, non sine coelesti aura, adsciscentis et induentis, in uterum augenod illi idoneum, aptam commissionem." Ita censuit naturae haud incelebris indagator (*Traite des Animaux*, Part 3. Chap. derniere) *Claudius Perraultius Gallus*; ita etiam nuperius praestantissimus G. Wollastonus (*Religion of Nature delineated*, Sect. 5. Subsect. 15) nostras, ut scriptis suis prodidit uterque.

Neque enimvero novum prorsus aut hujus aevi est hoc placitum; namque sicuti *Pythagoras* cum sequacibus suis, solem stare, terram in circuitum ferri (quod ab *Evis* didicerat ille) statuebant, licet in calamitosam rerum coelestium scientiae perburbationem aliter senserint posteri; ita eximius inter sui temporis sophos *Anaxagoras*, uti de eo testatum reliquerunt *Romanorum* doctissimus Varro, nec indoctior inter Graecos Theophrastus, semina ex aere in terram prolabi docebat. Sic itidem ut profligata *Aristotelica* et *Ptolemaica* de coelis haeresi invaluit antiqua et vera de iis opinio asserente eam *Copernico*, ita etiam hanc de generatione hypothesin apud posteros invalituram esse rem totam serio expendenti, dubitandi vix superfore locum augurari licebit.

EXPERIMENTS AND CONSIDERATIONS ON THE GENERATION OF PLANTS, BY JAMES LOGAN, LONDON: PRINTED FOR C. DAVIS, OVER-AGAINST GRAY'S INN GATE, HOLBORN. 1747. Translated by Dr. John Fothergill.

The Translator's Advertisement.

The following Essay in *Latin* was published at *Leyden* in 1739: It is now translated and reprinted here, that the Sentiments contained in it may be submitted to more general Consideration. Our Author's Address in choosing and conducting Experiments, and his Capacity for the abstrusest Researches, would doubtless have enabled him to have given the World ample Satisfaction on this intricate Subject, had he been permitted to prosecute his Inquiries. But his Country called him to more important Affairs, and kept him constantly engaged in Employments more immediately beneficial to Society.

The Translator has endeavoured [to] keep close to his Author's Sense: In point of Expression, he fears, he often falls short of the Original, the Style whereof is nervous, concise, and truly *Roman*. The *Latin* botanical Terms are mostly retain'd; as we have not yet Words in our own Tongue sufficiently classical whereby to express the various Parts of Plants and Flowers, which the growing Science is obliged to describe; and to explain by Terms adopted from other Languages.

It is hoped this little Essay will be received with the Candour due to those who endeavour to please and instruct; if the Translator has failed in promoting these Ends, he can only plead his Intention to the contrary, and request the Indulgence both of the Public and his Author.

J. F.

As several Doubts had formerly occurr'd to me in respect to the Generation both of Plants and Animals, when I first heard of the *Farina foecundans*, or impregnating male Dust, I conceived great Hopes that these would be easily solved, and the Whole of this intricate Affair receive considerable Light from the Discovery. And as I had long ago observed, with Surprize, the singular Way of Growth of our *Indian* Wheat or Maize, I judged it, of all the Plants I had seen, or perhaps of any that Nature produces, the most proper one for Experiments of this kind.

Indian Wheat grows to the Height of 6, 8, and sometimes 10 Feet. At the Top of the Stalk it bears a thready Tuft or Tassel (called by *Malpighi Muscarium*)[18] furnish'd with *Apices*, which yield the

[18] Marcello Malpighi (1628–1694), an Italian botanist and anatomist, was the founder of microscopic plant anatomy, and contributed to the science of pollen morphology, as did his younger contemporary, Nehemiah Grew (1641–1712), although Grew was probably more important in this field. Malpighi's papers on botany, zoology, and medicine were published by the Royal Society of London in 1687 as *Opera omnia*. In his studies

Farina. From the Joints of the Stalk below, the Ears grow out, which are 6, 8, 10, and sometimes even 12 Inches long. These consist of a pretty solid Substance, about an Inch thick, set quite round with Grains regularly disposed in Rows, in a very beautiful Manner. Generally there are eight such Rows, often ten, sometimes twelve; and I once saw sixteen; There are commonly forty Grains in each Row, more or less; which, in their first Rudiments, and whilst the Stalk they grow upon is soft and tender, may justly be called the *Ova* or Eggs: To each *Ovum* there adheres a white, fine smooth Filament, which, excepting that it is hollow, resembles a Thread of Silk. These Filaments are disposed one by one in Order, betwixt the Rows from that End where the Ear rises from the Stalk to the Other, where they creep from under the Case that Incloses the Ear, and make their Appearance, in the open Air, in a Bundle or Skein: Their Colour in this Part is mostly whitish, though sometimes a little yellow, red, or purple, according to the Nature of the Plant they grow from: These Filaments, as I formerly suspected, are the real Styles of the Eggs.

Intending therefore to make some Experiments on this Plant, towards the End of *April*, I planted four or five Grains on Hillocks, as is usual in sowing Maize, in each Corner of a Little Garden I had in Town, which was about 40 Feet wide, and 80 long. About the Beginning of *August*, when the Plants were full-grown, and the tufts on the Top, and the Ears on the Stem, had acquir'd their full Extent, I cut off these Tufts from every Plant on one Hillock. On another, without meddling with the Tufts, I gently open'd the Leaves that cover'd the Ears, and cut away from some all the Styles, and then closed the Leaves again; from others a Quarter Part, from others one Half, and from others three Quarters, and left the rest untouched. I cover'd another Ear, before the Skein of Styles appear'd out of the Case with a Piece of very fine and soft Muslin, but so loosely, that its Growth could not be injur'd; and whilst the furzy Texture of the Muslin suffer'd it to receive all the Benefit of the Sun, Air, and Showers, the *Farina* was effectually secluded. I left the Plants on the fourth Hillock, as I did these, except in the Circumstances above-mentioned, unmolested, till they were fully ripe.

About the Beginning of *October*, when it was Time to inquire into the Success of my Experiments, I made the following Observations. In the first Hillock, where I had cut off all the Tufts, the Ears, whilst they remained covered with their Husks, look'd indeed very

well, but were small, and felt light when handled; and not one perfect Grain to be found in them, except in one large Ear, which grew out somewhat farther from the Stalk than usual, and on that Side too which faced another Hillock in a Quarter from whence our strongest Winds most commonly blow: In this Ear alone I found about twenty Grains which were full-grown and ripe. I attributed this to some *Farina* brought by the Wind from a distant Plant. In those Ears from which I had plucked off some of the Styles, I found just so many ripe Grains as I had left Styles untouched. In those covered with Muslin, not one ripe Grain was to be seen: The empty or barren Eggs were nothing but mere dry Husks.

From these Experiments, which I made with the utmost Care and Circumspection, as well as from those made by a great many other Persons, it is very plain, that this Farina, emitted from the Summits of the Styles, is the true Male Seed, and absolutely necessary to render the Uterus and Grain fertile. A Truth which, however, certain, yet was unknown till the present Age: The Discoverer of this grand Secret of Nature ought ever to be remember'd with due Applause. Sir Thomas Millington,[19] sometime Savilian Professor, seems first to have taken notice of it, before or about the Year 1676, according to the Account which Dr. Grewe gave, in a Lecture read before the Royal Society the 9th of November the same Year. (See Grewe's *Works*, p. 161, 171.)[20] Malpighi nowhere, that I know of, mentions its Use. And *Grew* himself, tho' he allows it necessary for Fecundation, yet did not suspect, that it enter'd the *Uterus*: But *S. Morland*, about twenty Years after, asserted, that it enter'd the *Uterus* thro' the Canal of the Style.[21]

---

[19] Sir Thomas Millington (1628–1704), a physician and one of the original members of the Royal Society of London, was the Sedleian Professor of Natural Philosophy at Oxford University from 1675 until his death. He was not a Savilian Professor, as Grew inaccurately described him in 1676. It is because of the passage from Grew cited by Logan that Millington acquired the reputation of having been the discoverer of sexuality in plants, but the credit probably belongs to Grew. See the biographical sketch of Millington in *The Dictionary of National Biography* **13**: p. 442.

[20] Logan's reference is to Nehemiah Grew, *The Anatomy of Plants* (1682). His copy is in the Library Company of Philadelphia. For a recent reprint see Nehemiah Grew, *The Anatomy of Plants with an Idea of a Philosophical History of Plants and Several Other Lectures Read before the Royal Society*, reprinted from the 1682 edition with a new introduction by Conway Zirkle (*The Sources of Science*, No. 11, New York and London, Johnson Reprint Corporation, 1965).

[21] Sam. Morland, "Some new Observations upon the parts and use of the Flower in Plants," Royal Society, *Philosophical Transactions* **23**, No. 287 (Sept. and Oct., 1703): pp. 1474–1479. Morland, who was probably influenced to some extent by Malpighi as well as Grew, suggested an "analogy between *Animal* and *Vegetable* Generation. . .." (*Ibid.*, p. 1479.) Morland could only suggest that the male seed, or farina, entered the vasculum seminale (uterus) through the canal of the style. He had not observed it, and recommended that observations be made with microscopes. (*Ibid.*, pp. 1475–1476.) According to

of plant anatomy, Malpighi "was obsessed with the idea that every part of the plant had its counterpart, both in form and function, in the animal body. . .." (R. P. Wodehouse, *Pollen Grains*, pp. 20–23.) Logan's use of zoological terms in designating the sexual parts and functions of plants indicates the influence of Malpighi. The word "muscarium" is the same as flabellum which is now more commonly used to indicate fan-shaped organs or structures of insects or crustacean life.

(See Phil. Trans. No. 287.) I once saw a small Grain in the Middle of this Canal; nor is it to be doubted, but that stricter Inquiries will discover more of them passing the same way.[22]

It is true, that what has thus far been said is now very well known to every one, who is in the least acquainted with these Matters; but the Observations that follow may not perhaps be unworthy the Attention of the Curious. Want of Health, and sufficient Leisure, after I had made these Experiments, prevented me from prosecuting them any farther; but, as these permitted, I put down the following Remarks.

Not only in this Plant, in Nut-bearing Trees, Reeds, in all the Tribe of Gourds, as Pompions, Melons, Cucumbers, &c. in which the Male and Female Parts of Generation are separately placed, but also in most of those Flowers, which, from both Parts being placed within the same Flower-cup, are by some called *Hermaphrodites*, the *Apices* are so situated, that, after the *Farina* is perfected, they can seldom if ever, touch the Summit of the Style, or *Os Uteri*. But in these, as well as in such where the Organs are separately placed, the *Farina* must of Necessity, after it is thrown off from the *Apices*, float in the circumambient Air, and be Subjected to the Hazard of not reaching the *Os Uteri*, and performing its Office there (tho' this is absolutely requisite for Fructification); but is left to the Mercy of the Wind and Rains, and may be wholly lost.

Of the same Nature appears to be the remarkable Difference observable in the *Apparatus* which Nature makes use of in forming and defending the *Uterus* and its Eggs on one Hand, and in providing the Chives,[23] the *Apices*, and *Farina*, on the other. The *Uterus* or *Ovarium* is produced from the Heart of the little Branch or Stalk it grows on, and is on all Sides strongly secured, according to the Nature of the Plant; excepting in a few of those Plants called *Gymnospermae*. At the same time the Eggs, till they are render'd fruitful by the Male Dust, seem only to consist of a thin Pellicle full of an homogeneous Liquid: Whilst the Chives, designed to bear this impregnating Substance, grow either out of the *Petala*, which soon fall off like Hairs from the Skin, or from the Parts whence the *Petala* spring; as may be observed in those Flowers whose Styles are protected by a Case or staminous Sheath, as in Mallows, &c. Which two, *viz.* the Style and its Sheath, tho' they seem to compose only one Body, yet arise from very different Parts. For the Style in every Plant grows from the Middle of the Ovary; but the *Vagina* adheres

to the *Petala*, being formed of Fibres detached from the same Origin with them. The *Apices* or Summits consist of Pellicles, filled as it appears, with a kind of homogeneous Pulp or *Parenchyma*; which, nevertheless, whilst it is soft and moist, may be observed to consist of little Grains. As the thin Coats of these *Apices* by degrees become dry, they burst, and the *Farina* is turn'd out; which at first sticks so fast to these Coats, that it cannot be easily shaken off; but, as the *Apices* hang loosely by a very slender Thread, and are constantly moved by the Air, the *Farina* gradually falls off, and sticks like Dust to various Parts of the Flowers, but especially to the Tops of the Styles, which are furnished with a viscid clammy Liquor, called by *Malpighi*, a *Turpentine*. But if it happens, that, for a few Days, there is not Wind enough to stir the *Apices*, if one shakes the Stalk of the Maize, a large Quantity of the *Farina* flies up like a Cloud of Smoke or Dust. There is likewise farther to be observed in the Maize, that, on the same Day when the *Apices* burst, and hang loosely waving in the Air, the Skein or Bundle of Styles appear from under the Husk or Sheath that covers the Ear, and are in like manner exposed. The Circumstance should put us upon observing what happens in this Respect to other Plants.

The Particles of the *Farina* have all the same Figure in the same Species of Plants; but are different in different Species. In most Plants they are round or globular; in many they are oblong, like Grains of Wheat; in some, as the Marigold, Mallow, and others, they appear like an indented Wheel, or as a Globule set round with Prickles. I never saw any angular ones, as *Grew* describes in the Pansy (*Anatom.* TAB.58.);[24] for the most part they are smooth and shining; in the Maize they seem flat.

But in the Generation of the Seed itself, and the Process of its Growth, as described by *Malpighi* and *Grew*, and which I have likewise observed, there are several Things which create Astonishment, and justly claim our serious Consideration.

In the first place, when all the gay Pomp of the Flower fades, the Styles drop off, and the Egg begins to swell, a little Bladder, which *Malpighi* calls the *Amnion*, appears at its Point; this is furnished with an umbilical Cord, which passes thro' the *Chorion* to the opposite Part of the Seed. The *Amnion* increasing with the Egg after a few Days, another little Body is discovered, which in like manner daily increases, till it fills up the whole *Amnion*, as the *Amnion* does the *Chorion*, and this at length the whole Egg. These two, *viz.* the *Amnion* and *Chorion*,[25] then become the

Wodehouse, this theory was stated again twenty years later by Christian Wolff (1723). (Wodehouse, *Pollen Grains*, p. 46.) Logan does not mention Wolff in this connection.

[22] This was a remarkable observation, because such a grain is so minute in size that it is extremely difficult for the human eye to see it. That Logan saw it at all testifies to the very great pains he took in his observations.

[23] Threads would seem to be a better translation than chives.

[24] Nehemiah Grew's *Anatomy of Seeds* appeared in 1677, and his *Anatomy of Plants* in 1682. The latter work consists of four "books"; the fourth book is "The Anatomy of Leaves, Flowers, Fruits, and Seeds" and contains 42 tables. This appears to be the work cited by Logan.

[25] The terms "amnion" and "chorion" are more commonly used today in zoology. The amnion is the foetal membrane of reptiles, birds, and mammals; it is also used to indicate the inner

Husks or Coats of the Seed; and this, when it has acquired its full Size and Maturity, is the real Seed: when the external Husk is removed, it is found to consist for the most part of two Lobes: Sometimes it is not divided at all, as in Bread-Corn.

In this Process it is particularly to be remarked, that, whilst the Seed is increasing, another small Body is observed, in some, at its Point, as in the nuciferous, pomiferous, and some other Trees; in others, at one Side; which, at first, looks like some adventitious Body, and foreign to the Seed, with which it appears to have scarce any Connexion. By degrees it puts out just so many little Veins as there are Lobes in the Seed: These Veins connect this Body to the Seed, and afterwards disperse themselves through the Lobes, as the Arteries are spread thro' the Leaves.

In those Seeds wherein this little Body is placed at their Point, it seems to be very distinct from them: It consists of two Parts; one of which, when the external Teguments are taken off, appears more conspicuous, and is called by *Grew* the *Radicle*, because it forms the Root of the Plant; to the other, which lies hid betwixt the Lobes of the Seed, he gives the name of *Plume*, because, as it contains the Embryo, Stalk, and Foliage of the future Plant, it has somewhat of the Appearance of a Feather: Both these Parts are united by a kind of slender Joint.[26]

In the larger Seeds, whose Lobes adhering to the Stalk are turned out of the Ground, and become seminal Leaves, this Plume is very conspicuous; in Kidney-Beans especially it is extremely beautiful. In Beans, Pease, and other Plants of this kind, whose Lobes do not rise out of the Ground, the Connexion of the Plume with the Lobes appears more firm and compact; but when the second Coat of the Seed, which is the first or external one of the Plume, is removed, it has just the same Appearance as in the nut-bearing and other Trees above-mentioned; that is, it looks as if it were a foreign or adventitious Body grafted upon the Seed. This in reality is the true Seed of the Plant, or rather the Plant itself, form'd and prepar'd for Vegetation against the following Season; whilst the rest of the Seed serves only for the Food of Animals, and to supply the tender Plant with proper Nourishment, till it is become strong enough to prepare and extract it from the Earth.

It is very certain, that the whole Plant is drawn or unfolded out of this little Seed in Miniature; save that, in some, the outer and inner Coats of the Seed give a Covering to the Stalk: For tho' in several Kinds of Corn, as Barley, Oats, and Maize, the Radicle on one Side of the Seed becomes the Root, and the Stalk shoots out on the opposite; yet this Stalk proceeds

from the same Radicle here, as well as in other Plants; for it passes from the Radicle in the Form of a Vein, or subcutaneous Canal, till it shows itself on the other Side, without appearing to communicate with the rest of the Seed.

Having thus enquired into the Process of Generation, let us now consider what Use may be made of our Observations. In the first place, we see that the *Farina* is absolutely necessary for rendering the *Uterus* or *Ova* fruitful: That to produce this *Farina*, Nature only makes use of a very simple *Apparatus*; that, when perfected, it is frequently committed to the Air and Wind, before it can reach the *Os Uteri* or Style: A process, which, as it differs so much from the common Conduct of Nature in other Cases, must doubtless be founded upon some substantial, tho' less obvious Reason. For, as it is manifest that Nature always makes choice of the shortest and most certain means to obtain her End; so it cannot be doubted, but that a more expeditious and compendious Method of conveying the *Farina* to the *Uterus* would have been employed, had not something of greater Consequence, which was to be regarded, prevented it. Besides, we find nothing in the *Ova* before Impregnation, except a kind of liquid Substance, called by *Malpighi Colliquamentum*.[27] In the *Amnion* a new little Substance shews itself, and makes up the Bulk of the Seed: Lastly, in this Seed the true seminal Principle, the real Plant itself, appears, as if it was something foreign, and grafted upon the other Seed.

If then we deliberately view the Whole of this Process, and consider the Structure of the Plant itself, and of how many and how different Parts it consists, namely, of a Root, a Stalk, Leaves, Canals, Strainers, and other Instruments without Number, if we likewise reflect, by what Means and Passages the Rudiments of so many different Parts are convey'd into this little Particle, by what Machinery they are formed and united into such amazing Order and Beauty, it is scarcely possible to conceive, that so wonderful a Structure should be the Effect of so simple a Process. It not only exceeds the utmost Limits of human Apprehension, as indeed a great many other Subjects do, but even seems absurd to Reason.

As therefore the Article of Generation has to this Time lain under insuperable Difficulties, and as all the Methods of accounting for it, hitherto invented, have at last been found defective, why may we not suppose, that Nature herself points out the whole Process of the Affair; especially in forming the seminal Substance, distinct from the rest of the Seed? namely, that the *Farina* is for this Purpose committed to the Air, that it may receive out of the Air this little Seed or Plant, praeexistent and completely formed, tho' in *Stamina* inconceivable minute and invisible; and thus become pregnant hereby? It is drawn by an inherent attrac-

---

embryonic membrane of insects. The chorion is an embryonic membrane external to and enclosing the amnion.

[26] In the botanic terminology of today the small body to which Logan refers is called the tigellum. The tigellum consists of radicle and plumule. Grew's "plume" was the plumule.

[27] The colliquament is defined in modern dictionaries as the first rudiments of an embryo.

tive Force (like as the Males and Females of Animals we see are mutually, and even at a Distance, affected) first into the Style, and thro' that it slides by proper Canals to the *Ova*: And from this *Farina*, nourished by the Juices of the Plant for the Purposes above described, the Bulk of the Seed is formed. Lastly, the little Plant hid in the Seed, and, cloathed with a terrestrial Matter, which it borrows from the *Farina*, exerts itself; and, increasing by proper Nutriment, which it draws from the Earth, at length springs up.

Nor perhaps is this Method of accounting for Generation to be confin'd to Vegetables only; but it may, with equal Reason, be applied to the Production of every living Creature that makes a Part of this Universe of Things. "Whatever in this World have Life (Plants as well as Animals) first of all exist completely formed in Miniature, or in their first Rudiments, and are taken from the Air or *AEther*; then, being cloathed with an earthy Substance, they increase and grow larger. Nor is Generation any thing else than the placing this Principle in the male Seed, where it acquires an earthy Nature, tho' not without some Portion of an *Aura coelestis*, and a proper Conveyance of it into a suitable *Matrix*, where it may grow and increase." This was the Opinion of the ingenious *Claude Perrault*[28] long ago, as well as of the celebrated Wollaston[29] our Countryman, as appears from the Writings of these great Men.

Nor is this Notion altogether new, or peculiar to the present Age. For, as *Pythagoras* and his Followers asserted, that the Sun stood still, and the Earth moved, (as he had learned from the Eastern Philosophers) tho' to the great Confusion of Astronomy, some who came after him thought otherwise: So *Anaxagoras*, who was justly eminent amongst the Philosophers of his Time, as the learned *Varro*[30] and no less celebrated *Theophrastus*[31] inform us, taught, that the Seeds of all Things were in the Air, and descended from thence. Thus, in like manner as *Copernicus* restored the true Doctrine in relation to the heavenly Bodies, and banished the Errors of *Aristotle* and *Ptolemy*, so there is just Room to apprehend that this Hypothesis concerning Generation will be readily adopted by Posterity.

---

[28] Claude Perrault (1613–1688) began adult life as a doctor of medicine, but changed his career and became most famous in his time as an architect. Besides his work on architecture, Perrault was the author of *Essais de Physique, ou recueil de plusieurs traitez touchant les choses naturelles* (4 v., Paris, 1680–1688) and *Memoires pour servir a l'histoire naturelle des animaux. . .* (1671).

[29] William Wollaston (1660–1724), *Religion of Nature Delineated* (1724).

[30] Varro *de re rustica*, lib. 1, c. 40. "In the first place, the seed, which is the origin of growth, is of two kinds, one being invisible, the other visible. There are invisible seed, if, as the naturalist Anaxagoras holds, they are in the air, and if the water which flows on the land carries them, as Theophrastus writes." [English translation by William Davis Hooper in the Loeb Classical Library edition of Marcus Porcius Cato, *On Agriculture*, and Marcus Terrentius Varro, *On Agriculture* (Cambridge: Harvard University Press, 1960), p. 267.]

[31] Theoph. *Hist. Plant.* lib. 3. cap. 2. Anaxagoras says that the air contains the seeds of all things, and that these, carried down by the rain, produce the plants. . .. Theophrastus, *Enquiry into Plants and Minor Works on Odours and Weather Signs*, trans. into English by Sir Arthur Holt (2 vol., Cambridge: Harvard University Press, 1961), I, 163. Logan's citation is in error. It should read lib. 3, cap. 1. Idem de causis plant. lib. 1. cap. 5. But if, in truth, the air also supplies seeds, picking them up and carrying them about, as Anaxagoras says, then this fact is much more likely to be the explanation; for it would produce other examples of beginnings and nourishings. [Robert Ewing Dengler, *Theophrastus: De Causis Plantarum Book One: Text, Critical Apparatus, Translation, and Commentary* (A Thesis in Greek, Philadelphia, University of Pennsylvania, 1927.)]

(This bibliography is limited to authors and titles which are relevant to the content of this volume. Logan, of course, owned many more scientific books, both ancient and modern. All titles listed below are preserved in the Library Company of Philadelphia.)

Acta eruditorum anno 1682–1731 publicata. Leipzig. v. 1–22, 24, 26, 28–31, 33–46 (1682–1703, 1705, 1707, 1709–12, 1714–1727) in 38 v. (James Logan's signature appears on the title page.)

ARCHIMEDES of Syracuse (287–212), Ἀρχιμήδουζ . . . *Archimedis Opera. Quae Extant. Novis demonstrationibus commentariisque illustrata. Per Davidem Rivaltum a Flurantia Coenomanum* . . . . Paris: for Claudius Morellus, 1615.

—— *Archimedis Opera: Apollonii Pergaei Conicorum Libri IIII. Theodosii Spaerica* . . . per Is. Barrow. London: William Godbid and sold by Rob. Scott. 1675.

—— Ἀρχιμήδουζ τοῦ Συρακουσίου, . . . *Archimedis Syracusani ac Geometrae excellentissimi Opera, quae quidem extant, omnia . . . Adiecta quoq. sunt Eutocii Ascalonitae in eosdem Archimedis* . . . Basel: Johann Heruagius, 1544.

—— *Des Unvergleichlichen Archimedis Kunst-Bucher, oder heutiges tages befindliche Schrifften* . . . *erlautert von Johanne Christophoro Sturmio* . . . . Nurnberg: Christoph Gerhard for Paulus Furstens, 1670.

—— *Des Unvergleichlicen Archimedis Sand-Rechnung. Oder Tieffinnige Erfindungeiner mit vermunderlicher Leichtigkeit augsprechlichen . . . erlautert von Joh. Christoph Sturm* . . . . Nurnberg: Christoph Gerhard for Paul Furstens, 1667.

ARGOLI, ANDREA (1570–1653), *Andreae Argoli Ephemeridum Iuxta Tychonis Hypotheses et Coelo deductas observationes Tomus Primus* [Tertius] *Ab Anno 1631 ad An 1655.* Padua: Frambotti, 1638. (Vol. III, Venice.) Logan used the astronomical tables in this book.

BACON, ROGER (1214?–1294), *Rogerii Bacconis Angli Viri eminentissimi Perspectiva nunc primum in lucem edita Opera et studio Johannis Combacchi* . . . . Frankfurt: for Anton Humm by Wolfgang Richter, 1614.

BARROW, ISAAC (1630–1677), *Lectiones Geometricae; In quibus* (*praesertim*) *Generalia Curvarum Linearum Symptomata Declarantur.* London: by Gulielmi Godbid for Johannem Dunmore, & Octavianum Pulleyn Juniorem, 1670.

—— *The Works of the Learned Isaac Barrow, D. D.* . . . . Published by the Reverend Dr. Tillotson, ... The first Volume. London: M. Flesher for Brabazon Aylmer, 1683–1687. 4 vols., London, 1700.

BOULLIAU, ISMAEL (1605–1694), *Ismaelis Bullialdi. Astronomia Philolaica. Opus novum, In quo motus Planetarum per novam ac veram Hypothesim demonstrantur. Mediique motus . . . Superque illa Hypothesi Tabulae . . . Addita est nova methodus cuius ope Eclipses Solares* . . . . Paris: for Simeon Piget, 1645.

BRADLEY, RICHARD (1688–1732), *New Improvements of Planting and Gardening, both Philosophical and Practical* . . . *The Second Edition Corrected* . . . [Third Part]. London: Printed for W. Mears, 1718. Logan also owned the 1717, 1724, 1730–31, and 1739 editions.

—— *The Artificial Gardener. The Second and Last Part. Containing. I. The Nature of the Hot-Bed* . . . . IV. *A New Method of preserving Exotick Plants in the severest Seasons.* London: for E. Curel, 1717.

—— *A Complete Body of Husbandry . . . with particular Directions for the fertilizing of Broom Ground* . . . . London: for James Woodman and David Lyon, 1727.

—— *The Country Gentleman and Farmer's Monthly Director* . . . . London: for James Woodman and James Lyon, 1726. Logan also owned the 1727 edition published in Dublin by S. Powell for George Ewing.

—— *The Country Housewife and Lady's Director, in the Management of a House, and the Delights and Profits of a Farm, Containing Instructions for managing the Brew-House* . . . . London: for Woodman and Lyon, 1727.

—— *Dictionarium Botanicum: Or, a Botanical Dictionary for the use of the curious in husbandry and gardening* . . . 2 vols., London: for T. Woodward and J. Peele, 1728.

—— *A General Treatise of Husbandy and Gardening; Containing a New System of Vegetables* . . . *In Two Volumes* . . . *Adorn'd with Cuts.* London: for T. Woodward and J. Peele, 1726.

—— *The Gentleman and Farmer's Guide, for the Improvement of Cattle* . . . *also the best Manner of Breeding, and Breaking Horses, both for Sport and Burden; with an Account of their respective Distempers,* . . . *Illustrated with Copper Plates.* London: J. Applebee, for W. Mears, 1729.

—— *The Gentleman and Gardners Kalendar, Directing What is necessary to be done every Month* . . . *To which is added, The Design of a Green-House* . . . . London: Printed for W. Mears, 1718.

—— *New Experiments and Observations, Relating to the Generation of Plants* . . . *Together with an Account of the extraordinary Vegetation of Peaches* . . . . London: for T. Corbett, 1724.

—— *A Philosophical Account of the Works of Nature.* . . . *Mineral, Vegetable, & Animal Parts of the Creation . . . Account of the State of Gardening . . . the Propagating of Timber-Trees, Fruit-Trees, &c.* London: for W. Mears, 1721.

—— *A Philosophical Enquiry into the late severe winter, the scarcity and dearness of provisions, and the occasion of the distemper . . . with letters from . . . physicians* . . . . London: for J. Roberts, R. Montagu and booksellers, 1729.

—— *The Riches of a Hop-Garden explain'd, From the several Improvements arising by that Beneficial Plant* . . . . London: for Charles Davis and Thomas Green, 1729.

—— *A Survey of the Ancient Husbandry and Gardening, collected from Cato, Varro* . . . . London: for B. Motte, 1725.

—— *Ten Practical Discourses Concerning Earth and Water, Fire and Air, As they relate to the Growth of Plants. With a Collection of New Discoveries for the Improvement of Land, either in the Farm or Garden* . . . . Westminster: Printed by J. Cluer and A. Campbell, for B. Creake, 1727.

BRAHE, TYCHO (1546–1601), *Tychonis Brahe. Astronomiae Instrauratae Mechanica.* Nuremberg: for Levinus Hulsius, 1602.

—— *Tychonis Brahe Epistolarum Astronomicarum Libri* . . . . Uranienburg: by the author, 1596.

—— *Tychonis Brahe Mathim: Eminent: Dani Opera Omnia, sive Astronomiae Instauratae Progymnasmata In duas partes distributa* . . . . Frankfurt: Johann Godofred Schonwetter, 1648.

CHALES, CLAUDE FRANÇOIS MILLET DE (1611–1678), *R. P. Claudii Francisci Milliet De Chales Camberiensis . . . Cursus seu Mundus Mathematicus Tomus Primus* [-Quartus] . . . 4 vols., Lyon: Anissoniana, Jean Poseul and Claude Rigaud, 1690.

COPERNICUS, NICOLAS (1473–1543), *Astronomia Instaurata.* Amsterdam, 1617.

CROUSAZ, JEAN-PIERRE DE (1663–1748), *Traité du Beau. Ou l'on montre en quoi consiste ce que l'on nomme ainsi* . . . . Amsterdam: François L'Honore, 1715.

—— *Commentaire sur l'Analyse des Infiniment Petits.* Paris: Laurent Rondet for Montalant, 1721.

DESCARTES, RENÉ (1596–1650), *Geometria, a Renato Des Cartes Anno 1637 Gallice edita; . . . Una cum Notis Florimondi de Beaune . . .* Amsterdam: Ludovic and Daniel Elzevir, 1659. Second title: *Principia Matheseos Universalis . . .* 1661. Logan also owned a copy of the third edition, published in 1683.

—— *Meditationes De Prima Philosophia . . . Editio Ultima . . .* (Second title:) *Appendix continens Objectiones Quintas & Septimas . . .* (Third title:) *Epistola Renati Des Cartes Ad . . . D. Gisbertum Voetium.* Amsterdam: Daniel Elzevir, 1678.

—— *Passiones Animae Per Renatum Des Gartes:* Gallice ab ipso conscriptae, nunc autem in exterorum gratiam Latina civitate donata Ab H. D. M. I. V. L. Amsterdam: for Ludwig Elzevir, 1650. Logan also owned a copy of the 1672 edition.

—— *Principia philosophiae . . . .* Amsterdam: Daniel Elzevir, 1672.

—— *Renati Descartes Epistolae, Partim ab Auctore Latino sermone conscriptae, partim ex Gallico translatae.* Amsterdam: Daniel Elzevir, 1668. Logan also owned a copy of the 1682–83 edition published by Blaviana in Amsterdam.

—— *Renati Des Cartes specimina Philosophiae: seu dissertatio de methodo Recte regendae rationis, & veritatis in scientiis investigandae: Dioptrice, et Meteora . . .* Ultima editio . . . . Amsterdam: Daniel Elzevir, 1672.

DU HAMEL, Jean Baptiste (1624–1706), *Regiae Scientiarum Academiae Historia, in qua praeter ipsius Academiae originem & progressus . . . .* Paris: Estienne Michallet, 1698. Logan also owned the 1700 edition published by Thomas Fritsch in Leipzig, and the 1701 edition published by Jean Baptiste Delespine in Paris.

EUCLID (fl. ca. 300 B.C.), *Adiecta praefectiuncula in qua de disciplinis mathematicis nonnihil.* Basle: for Johann Hervagius, 1533.

—— *Euclidis Elementorum Geometricorum Libri Tredecim ex Traditione Doctissime Nasiri Dini Tusini Nunc primum Arabice impressi.* Rome: Typographia Medicea, 1594.

—— *Euclidis Elementorum Libri XV. Accessit Liber XVI . . .* Auctore Christophoro Clavio . . . . Frankfurt: Nicolaus Hoffman for Jonaeus Rhodii, 1607.

—— *Euclidis Elementorum Libri XV. breviter demonstrati, Opera Is. Barrow, ...* London: J. Redmayne for J. Williams and J. Dunmore, 1678.

—— *Elementorum Libri XV. Campani Galli Transalpini. In Eosdem Commentarii. Theonis Alexandrini in XIII Priores, Zamberto Veneto Interpreto. Hypsiclis Alexandrini, In Duos Posteriores, eodem Interprete.* Paris: Henri Estienne, 1516.

—— *Euclidis Elementorum Libri XV. Graece & Latine.* Paris: Gulielme Cavellat, 1557.

—— *The Elements of Euclid Explain'd, In a New, but most Easie Method: Together with the Use of every Proposition through all parts of the Mathematicks . . . Now made English . . .* The Fourth Edition. Oxford: L. Lichfield for W. Freeman, J. Knapton and W. Keble, 1704.

—— *Euclidis Quae Supersunt Omnia. Ex Recensione Davidis Gregorii . . . .* Oxford: At the Sheldon Theatre, 1703.

FLAMSTEED, JOHN (1646–1719), *Atlas Coelestis.* London, 1729.

—— *The Doctrine of the Sphere, Grounded on the Motion of the Earth, And the Antient Pythagorean or Copernican System of the World.* In Two Parts. London: Printed by A Godbid and J. Playford, 1680.

—— *Historiae Coelestis Britannicae. Complectens Stellarum Fixarum nec non Planetarum Omnium Observationes . . .* 3 vols., London: H. Meere, 1725.

GALILEI, GALILEO (1564–1642), *Discorso e Dimostrazioni Matematiche, intorno a due nuove scienze Attenenti alla Mecanica & i Movimenti Locali . . . .* Leyden: Elzevir, 1638.

—— *Siderius Nuncius, Magna Longeque Admirabilia Spectacula pandens, . . . Apprime vero in Quatuor Planetis Circa Jovis Stellam . . . atque Medicea Sidera Nuncupandos decrevit.* [2nd title.] Joannis Kepleri . . . *Dioptrice: seu Demonstratio . . . .* London: James Flesher for Cornelius Bee, 1653.

—— *Systema Cosmicum, Autore Galilaeo Galilaei Lynceo, Academiae Pisanae Mathematico extraordinario, . . . In quo quatuor dialogia, De Duobus Maximis Mundi Systematibus . . . .* London: sold by Thomas Dicas, 1663.

GASSENDI, PIERRE (1592–1655), *Institutio Astronomica.* The Hague, 1656.

GREGORY, DAVID (1661–1708), *Astronomiae Physicae & Geometricae Elementa . . . .* Oxford: Sheldonian Theatre, 1702.

—— *Catoptricae et Dioptricae Sphaericae Elementa . . .* Oxford: Sheldon Theatre, 1695.

GREW, NEHEMIAH (1641–1712), *Anatomie des Plantes qui Contient une Description exacte de leurs parties & de leurs usages & qui fait voir comment elles se forment, & comment elles croissent . . . .* Paris: Lambert Roulland, 1675.

—— *The Anatomy of Plants. With an Idea of a Philosophical History of Plants . . . .* London: W. Rawlins for the author, 1682.

—— *The Anatomy of Vegetables Begun. With a general account of vegetation founded thereon.* London: for Spencer Hickman, 1672.

HALES, STEPHEN (1677–1761), *La Statique des Vegetaux, et L'Analyse de L'Air . . .* Ouvrage traduit de l'Anglois, par M. Buffon . . . . Paris: Chez Debure l'ainé, by Jacques Vincent, 1735.

—— *Vegetable Staticks: Or, An Account of some Statical Experiments on the Sap in Vegetables: Being an Essay towards a Natural History of Vegetation . . . .* London: for W. & J. Innys et al., 1727.

HALLEY, EDMUND (1656–1742), *Atlas Maritimus & Commercialis; Or, a General View of the World, so far as relates to trade and navigation: describing all the coasts, ports . . . .* 2 vols., London: for James and John Knapton et al., 1728.

—— *Catalogus stallarum Australium sive Supplementum Catalogi Tychonici . . . .* London: Thomas James for R. Harford, 1679.

—— *A Description of the Passage of the Shadow of the Moon over England, in the total eclipse of the Sun, on the 22d day of April 1715 in the morning.* London: sold by J. Senex and William Taylor, 1715.

—— *A Description of the Passage of the Shadow of the Moon over England in the total eclipse of the sun on the 11th day of May, 1724 in the evening. Together with the . . . last total eclipse of 1715.* London: engraved and sold by John Senex, 1724.

—— [*Edmundi Halleii Astronomi dum viveret regii tabulae astronomicae.*] London: William Innys, before 1721. Unpublished proof sheets purchased by Logan in 1721, with ms. additions and notes by him.

HARRIS, JOHN (1666?–1719), *Astronomical Dialogues Between a Gentleman and a Lady: wherein the Doctrine of the Sphere, Uses of the Globes, And the Elements of Astronomy and Geography are Explain'd; . . .* The Second Edition . . . . London: for A. Bettesworth and J. Batley, 1729. Logan also had a copy of the 1745 edition, corrected by George Gordon.

—— *The Description and Uses of the Celestial and Terrestrial Globes; And of Collins's Pocket-Quadrant . . .* London: printed for D. Midwinter, 1713. Logan also had a copy of the sixth edition, published in 1725.

—— *The Description and Use of the Globes, and the Orrery. To which is prefixed, By way of Introduction, a Brief Account of the Solar System . . . .* London: for Thomas Wright, and Richard Cushee, 1731. Logan also had copies of the second edition, 1732, third edition, 1734, and the fifth edition, 1740.

—— *Lexicon Technicum: or, An Universal English Dictionary of Arts and Sciences: explaining not only the terms of art, but the arts themselves.* London: for Daniel Brown et al., 1704. Logan also had a copy of the second edition, published in 1708, and a copy of the third edition, published in 1716 (vol. I) and 1723 (vol. II).

HEVELIUS, JOHANN (1611–1687), *Johannis Hevelii Prodromus Astronomiae, Exhibens Fundamenta, quae tam ad novum plane & correctionem Stellarum Fixarum Catalogum construendum,*

. . . *Accessit Corollarii loco Tabula Motus Lunae Libratorii*
. . . Danzig: Johann Zachary Stoll, 1690.

HORROCKS, JEREMIAH (1617?–1641), *Jeremiae Horroccii, Liver Poliensis Angli, ex Palatinatu Lancastriae, Opera Posthuma; viz. Astronomia Kepleriana . . . Accedunt Guilielmi Crabtraeia . . . Observationes Coelestes. In calce adjiciuntur Johannis Flamstedii* . . .. London: William Godbid for J. Martyn, 1673.

HUYGENS, CHRISTIAAN (1629–1695), *Christiani Hugenii Zulichemii, Const. F. Horologium Oscillatorium, sive De Motu Pendolorum ad Horologia aptato Demonstrationes Geometricae.* Paris: F. Muguet, 1673.

—— *Const. F. Theoremata de Quadratura Hyperboles, Ellipsis et Circuli* . . .. Leyden: Elseviriana, 1651.

——Κοφμοδεωρος, *sive De Terris Coelestibus, earamque ornatu* . . .. The Hague: Adrian Moetjens, 1699.

—— *Libellus de Ratiociniis in Ludo Aleae. Or, The Value of all Chances* . . .. London: S. Keimer for T. Woodward, 1714.

—— *Chr. Hugenii opuscula posthuma* . . .. Leyden: Boutesteyn, 1703.

—— *Traité de la Lumiere. Ou sont expliqueés les causes de ce qui luy arrivé dans la Reflexion, & dans la Refraction* . . .. Leyden: Pierre Vander Aa, 1690.

—— *Zelhemii Toparchae. Opera reliqua. Opuscula Posthuma* . . . Amsterdam: Jansson-Waesberger, 1728. Two volumes in one.

KEPLER, JOHANN (1571–1630), *De Cometis Libelli Tres I. Astronomicus* . . .. Augsburg: Andreas Apergeri for Sebastian Mylius, 1619.

—— *Epitome Astronomiae Copernicae Usitata forma Quaestionum & Responsionum conscripta, inque VII. Libros digesta, quorum Tres hi priores sunt De Doctrina Sphaerica* . . .. Frankfurt: Johann Godefrid Schonwetter for Johan-Friderick Weissius, 1635.

—— *Tabulae Rudolphinae.* London, 1675. This is not a copy Logan owned, but he referred to this work in his correspondence.

LA HIRE, PHILIPPE DE (1640–1718), *New Elements of Conick Sections: Together with a Method for their Description on a Plane* . . .. London: for Dan. Midwinter, 1704.

—— *Sectiones Conicae. In novem libros distributae, In quibus quidquid hactenus observatione dignum cum a vereribus, tum a recentioribus Geometris traditum est, . . . Adjecta demum est Brevis Expositio Propositionum septem Librorum Conicorum Apollonii Pergaei* . . .. Paris: for Estienne Michallet, 1685.

LEEUWENHOEK, ANTON VAN (1632–1732), *Antonii a Leeuwenhoek, Continuato Arcanorum Naturae* . . .. Delft: Henry A. Kroonevelt, 1697.

LEIBNIZ, GOTTFRIED WILHELM (1646–1716), *Recueil de Diverses Pieces, sur la Philosophie, la Religion Naturelle, l'Histoire, les Mathematiques, &c. Par Messrs. Leibniz, Clarke, Newton, &* autres Autheurs célèbres. 2 vols., Amsterdam, 1720.

—— *Virorum celeberr. Got. Gul. Leibnitii et Johann Bernoulli Commercium Philosophicum et Mathematicum.* 2 vols., Laussane and Geneva, for Marcus-Michel Bousquet, 1745.

LINNAEUS, CAROLUS (1707–1778), *Amoenitates Academicae, Seu Dissertationes Variae Physicae, Medicae, Botanicae . . . Accedit Hypothesis Nova de Febrium Intermittentium Causa.* 6 vols., Leyden, 1749–64. Logan, of course, could not have owned all of these volumes during his lifetime.

—— *Bibliotheca Botanica recensens Libros plus mille de plantis . . . Editionis Loco, Tempore, Forma Lingua . . . Fundamenta Botanica* . . .. Amsterdam: Salomon Schouten, 1736.

—— *Classes Plantarum. Seu Systemata Plantarum Omnia a fructificatione desumta . . . Fundamentorum Botanicorum.* Leyden: Conrad Wishoff, 1738.

—— *Corollarium Generum Platarum, exhibens genera plantarum sexaginta . . . Accedit Methodus Sexualis.* Leyden: Conrad Wishoff, 1737.

—— *Fauna Suecica Sistens Animalia Sueciae Regni: quadrupedia, aves, amphibia, pisces, insecta, vermes* . . .. Stockholm: Laurent Salvius, 1746.

—— *Flora Zeylanica Sistens Plantas Indicas Zeylonae Insulae; Quae Olim 1670–1677. lectae fuere a Paulo Hermanno* . . .. Stockholm: for Laurentius Salvius, 1747.

—— *Fundamenta Botanica, in quibus Theoria Botanices Aphoristice traditur.* Editio tertia . . .. Amsterdam: Salomon Schouten, 1741.

—— *Genera Plantarum Eorumque Characteres Naturales* . . . Editio secunda aucta & emendata. Leyden: Conrad Wishoff and Georg. Jac. Wishoff, 1742.

—— *Genera Plantarum Eorumque Characteres naturales, secundum Numerum Figuram, Situm, & Proportionen omnium Fructificationis partium* . . . Editio secunda, nominibus Plantarum Gallicis locupletata. Paris: Sumpribus Michaelis Antonii David, 1743.

—— *Oratio de Necessitate Peregrinationum intra patriam . . . Accedunt Johannis Browalii examen epicriseos Siegesbeckianae . . . et Johannis Gesneri . . . Dissertationes* . . .. Leyden: Cornelius Haak, 1743.

MALPIGHI, MARCELLO (1628–1694), *Anatome Plantarum* . . .. 2 vols., London: John Martyn, 1675–1679.

—— *Dissertatio Epistolica de Formatione Pulli in Ovo* . . .. London: John Martyn, 1673.

—— *Dissertatio Epistolica de Bombyce* . . .. London: John Martin & James Allestry, 1669.

—— *Opera Omnia . . . Tomis Duobus comprensa* . . .. 2 vols., London: for Thomas Sawbridge (Vol. I), M. F. for R. Littlebury *et al.* (Vol. II), 1686.

—— *Opera Posthuma . . . Quibus praefixa est Ejusdam Vita a Seipso Scripta.* London: for A. & J. Churchill and Nigri Cygni, 1697.

—— *de Structura Glandularum Conglobatarum Consimiliumque Partium, Epistola, Regiae Societati Londini* . . .. London: for Richard Chiswell, 1697.

—— *de Viscerum Structura Exercitatio Anatomica . . . Accedit Dissertatio ejusdam de Polypo Cordis.* London: T. R. for Jo. Martyn, 1669.

MERSENNE, MARIN (1588–1648), *Harmonicorum Libri XII. In quibus agitur de sonorum natura, causis et effectibus: de consonantiis, dissonantiis, rationibus, generibus, modis, cantibus* . . . Editio Nova . . .. Paris: Thomas Jolly, 1652.

—— *Minimi Cogitata Physico Mathematica* . . .. 2 vols., Paris: for Antony Bertier, 1644.

MILLER, PHILLIP (1691–1771), *The Gardeners and Florists Dictionary: or a Complete System of Horticulture:* containing I. The Culture of a Kitchen Garden . . . Vol. II . . .. London: H. P. for Charles Rivington, 1724. Lacks Vol. I.

—— *The Gardeners Dictionary: Containing the Methods of Cultivating and Improving the Kitchen, Fruit and Flower-Garden . . . and Vineyard* . . .. Abridg'd from the Folio Edition, by the Author . . . in Two Volumes . . .. London: Printed for the Author, and sold by C. Rivington, 1735. Logan also had copies of the 1731 edition of Vol. I and the 1739 edition of Vol. II.

MOLYNEUX, WILLIAM (1656–1698), *Dioptrica Nova. A Treatise of Dioptricks, In Two Parts.* London: for Benj. Tooke, 1692.

MOORE, JONAS (1617–1679), *A New Systeme of the Mathematicks: Containing I. Arithmetick, as well Natural and Decimal, . . . VIII. A New Geography,* . . .. 2 vols., London: A. Godbid and J. Playford for Robert Scott, 1681.

NEWTON, ISAAC (1642–1727), *Analysis Per Quantitatum Series, Fluxiones, ac Differentias: cum Enumeratione Linearum Tertii Ordinis.* London: Pearsons, 1711.

—— *Arithmetica Universalis: sive De Compisitione et Resolutione Arithmetica Liber. Cui accessit Halleiani Aequationum Radices Arithmetice inveniendi methodus* . . .. Cambridge: for Benj. Tooke, 1707, 2nd edition, 1722. Logan also had the 1732 edition published in Leyden by Joh. and Herm. Verbeek.

—— *Lectiones Opticae. Annis MDCLXIX, MDCLXX, MDCLXXI* . . .. London: William Innys, 1729.

—— *The Mathematical Principles of Natural Philosophy . . . Translated into English by Andrew Motte. To which are added,*

The Laws of the Moon's Motion, according to Gravity. By John Machin . . . . 2 vols., London: for Benjamin Motte, 1729.

—— *The Method of Fluxions and infinite series* . . . Translated from the Author's Latin Original . . . By John Colson . . . . London: Henry Woodfall and sold by John Nourse, 1736.

—— *La Methode des Fluxions, et des Suites Infinies* . . . . Paris: De Bure, the elder, 1740.

—— *De Mundi Systemate Liber Isaaci Newtoni. Opus diu integris fuis partibus desideratum* . . . . London: J. Tonson, J. Osborn & T. Longman, T. Ward & E. Wicksteed, & F. Gyles, 1731.

—— *Neutoni Genesis Curvarum per Umbras. Seu perspectivae universalis Elementa* . . . . London: A. Millar, 1746.

—— *A New and most Accurate Theory of the Moon's Motion* . . . *And Published in Latin by Mr. David Gregory in his Excellent Astronomy.* London: A. Baldwin, 1702.

—— *Optical Lectures read in the Publick Schools of the University of Cambridge, Anno Domini, 1669. By the late Sir Isaac Newton* . . . *Never before printed. Translated into English out of the Original Latin.* London: for Francis Fayram, 1728.

—— *Optice: Sive de Reflexionibus, Refractionibus, Inflexionibus & Coloribus Lucis, Libri Tres* . . . . Second edition. London: William Bowyer for William and John Innys, 1719. Logan also owned the 1740 edition published at Lausanne and Geneva by Marc-Michael Bosquet and Co.

—— *Optices Libri Tres: Accedunt Ejusdem Lectiones Opticae, et Opuscula Omnia ad lucem & Colores pertinentia* . . . . Patavii: Typis Seminarii. Joannem Manfre, 1749.

—— *Optics: or, a Treatise of the Reflections, Refractions, Inflexions and Colours of Light. Also Two Treatises of the Species and Magnitude of Curvilinear Figures.* London: for Sam. Smith and Benj. Walford, 1704. Logan also owned a copy of the 1706 edition in Latin. Logan owned a copy of the third edition, 1721, printed in London for William and John Innys, and a copy of the fourth edition, corrected, printed in 1730.

—— *Philosophiae Naturalis Principia Mathematica* . . . . London: for the Royal Society by Joseph Streater, 1687. Logan also owned the 1714 edition published in Amsterdam; the 1723 edition published in Amsterdam with *Analysis per Quantitatum Series, Fluxiones ac Differentias cum enumeratione Linearum Tertii Ordinis;* the 1726 edition published in London; and a three volume edition published in 1739, 1740, and 1742 by Barrillot & Filii in Geneva.

—— *Sir Isaac Newton's two treatises of the Quadrature of curves, and analysis by Equations of an infinite number of terms, explained* . . . . London: by James Bettenham, for John Nourse, & John Whiston, 1745.

—— *Traité d'Optique sur les Reflexions, Refractions* . . . *de la Lumiere.* Traduit par M. Coste, dur la seconde Edition Angloise . . . Second Edition Françoise . . . . Paris: for Montalant, 1722.

—— *A Treatise of the System of the World* . . . Translated into English. London: for F. Fayram, 1728. Logan also owned a copy of the second edition, published in 1731.

—— *Universal Arithmetick: or, a Treatise of Arithmetical Composition and Resolution* . . . . 2nd edition, London: T. Wood for J. Senex *et al.*, 1728.

PERRAULT, CLAUDE (1613–1688), *Oeuvres diverses de Physique et de Mechanique,* de Messrs. C. & P. Perrault . . . Divisees en deux volumes. 2 vols., Leyden: Pierre Vander Aa, 1721.

PORTA, GIOVANNI BAPTISTA DELLA, of Naples (1540–1615), *Magiae Naturalis Libri Viginti, In Quibus Scientiarum* . . . *Accessit Index* . . . . Frankfurt: Samuel Hempel for Claude Marn and the heirs of Johann Aubrius, 1607. Logan also owned a copy of the 1651 edition published in Leyden for Peter Leffen.

—— *Natural Magick.* London: for Thomas Young and Samuel Speed, 1658. Two copies, one imperfect.

PTOLEMY, CLAUDIUS (fl. 127–151), *Almagestu Cl. Ptolemei Pheludiensis Alexandrini Astronom principis: Opus ingens ac nobile omnes celoru motus continens. Felicibus Astris eat in lucez.* Venice: Petrus Liechtenstein, 1515.

—— *Claudii Ptolemai Harmonicorum Libri Tres* . . . Johannes Wallis . . . . Oxford: Sheldon Theatre, 1682.

—— *Claudii Ptolemaei Pelusiensis Alexandrini Omnia, quae extant, opera, geographia excepta* . . . . Basel: Henric Petri, 1541.

—— *Claudii Ptolemaei Magnae Constructionis, Id est Perfectae coelestium motuum pertractationis, Lib. XIII. Theonis Alexandrini in eosdem Commentariorum Lib. XI.* Basel: Johann Walder, 1538.

—— *Claudii Ptolemai Tabulae geographicae Orbis Terrarum Veteribus cogniti.* Utrecht: for Francis Halman, Willem vande Water, and Franccker, for Leonard Strick, 1698.

—— *Geographia universalis uetus & nova, complectens Claudii Ptolemaei Alexandrini enerrationis libros VIII.* Basel: Henric Petri, 1547.

—— *Theatrum Geographiae Veteris, Duobus Tomis distinctum, Edente Petro Bertio Bevero* . . . . Amsterdam: Isaac Elzevir, Leyden, for Judocus Hondius, 1619.

RICCIOLI, GIOVANNI BATTISTA (1598–1671), *Astronomiae Reformatae. Tomi Duo, quorum prior observationes, hypotheses, et fundamenta tabularum,* . . . . Bologna: Heirs of Victor Benat, 1665.

SNELL, WILLEBRORD (1591–1626), *Willebrordi Snelli R. F. de re nummaria liber singularis.* Antwerp: Plantiniana, by Raphelengi, 1613.

STREET, THOMAS (*ca.* 1700), *Astronomia Carolina. A New Theorie of the Coelestial Motions* . . . . London: for Lodowick Lloyd, 1661. Logan also owned copies of the second edition published in London in 1710, and the third edition published in London in 1716.

THEOPHRASTUS (*ca.* 372 B.C.–*ca.* 287 B.C.), *Theophrasti Eresii de Historia Plantarum Libri Decem, Graece & Latine. In quibus Textum Graecum* . . . *illustravit Ioannes Bodaeus a Stapel* . . . *Accesserunt Iulii Caesaris Scaligeri,* . . . . Amsterdam: Judoco Broerssen for Henrik Laurent, 1644.

VARRO, M. TERENTIUS (116 B.C.–26 B.C.), *M. Terentii Varronis Opera Omnia Quae Extant. Cum Notis Iosephi Scaligeri,* . . . *His accedunt Tabulae Naufragii, seu Fragmenta* . . . . Durdrecht: Johann Berewont, 1619.

VITALI, GIROLAMO (Hieronymus Vitalis) (*ca.* 1670), *Lexicon Mathematicum Astronomicum Geometricum, Hoc est rerum omnium ad utramque immo & ad omnem fere mathesim* . . . *Adjecta brevi novorum theorematum expensione,* . . . *Disciplinarum omnium mathematicarum summa,* . . . . Paris: Ludovic Billaine, 1668.

—— *Absolutissimae Primi Mobilis Tabulae, ad integrum Quadrantum ex Triangulorum rationcinio concinnatae* . . . . Nuremberg: for the heirs of Wolfgang Mauritius Endteri and Johann Andreas Endteri, 1676.

VITRUVIUS, MARCUS POLLIO (fl. *ca.* 10 A.D.), *M. Vitruvii Pollionis De Architectura Libri Decem.* Amsterdam: Ludwig Elzevir, 1649.

VOSSIUS, GERHARD JOHANN (1577–1649), *Gerardi Ioannis Vossii De Quatuor Artibus Popularibus, de Philologia, et Scientiis Mathematicis, cui Operi subjungitur, Chronologia Mathematicorum, Libri Tres.* Amsterdam: Johann Blaev, 1660. Logan also owned a copy of the 1650 edition, in three volumes.

WALLIS, JOHN (1616–1703), *Mechanica: Sive, De Motu, Tractatus Geometricus* . . . . London: William Godbid, for Mosis Pitt, 1670.

—— *Opera mathematica.* 3 vols., Oxford: Sheldonian Theatre, 1693, 1695, 1699.

—— *Opera mathematicorum pars prima* . . . . Oxford: Leon. Lichfield for Tho. Robinson, 1657. Logan also had the 1656 edition.

—— *Tractatus Duo, Prior, De Cycloide* . . . *De Cissoide* . . . *De Curvarum* . . . . Oxford: Lichfield, 1659.

WARD, SETH (1617–1689), *Astronomia Geometrica: ubi methodus proponitur qua primariorum . . . Opus, Astronomis hactenus desideratum . . . .* London: Jacob Flesher, 1656.

WHISTON, WILLIAM, *Praelectiones Astronomicae Cantabrigiae in Scholis Publicis Habitae . . . Quibus Accedunt Tabulae Plurimae Astronomicae . . . .* Cambridge: At the University Press for Benj. Tooke, London, 1707.

WHISTON, WILLIAM (1667–1752), and HUMPHRY DITTON, *A New Method for Discovering the Longitude both at Sea and Land . . . .* London: for John Phillips, 1714. Logan also owned a copy of the second edition, published in 1715.

WOLFF, CHRISTIAN VON (1679–1754), *Elementa Matheseos Universea, . . . qui Commentationem de Methodo Mathematica, Arithmeticam, Geometriam, ...* 2 vols., Halle in Sachen: Rengeriana, 1713–1715. Logan also owned the 1717 edition, and the 1732 edition published in Geneva by Marais-Michael Bousquet and Co.

WOLLASTON, WILLIAM (1660–1724), *The Religion of Nature Delineated.* London: S. Palmer and sold by B. Lintott, W. and J. Innys *et al.*, 1725. Logan also owned a copy of the 1726 edition.

INDEX

MEMOIRS

OF THE

AMERICAN PHILOSOPHICAL SOCIETY

———————

TRANSACTIONS

OF THE

AMERICAN PHILOSOPHICAL SOCIETY
